Fundamentals of Intelligent Transportation Systems Planning

For a listing of recent titles in the *Artech House ITS Library,* turn to the back of this book.

Fundamentals of Intelligent Transportation Systems Planning

Mashrur A. Chowdhury
Adel Sadek

Artech House
Boston • London
www.artechhouse.com

Library of Congress Cataloging-in-Publication Data
A catalog record for this book is available from the Library of Congress.

British Library Cataloguing in Publication Data
A catalog record of this book is available from the British Library.

Cover design by Yekaterina Ratner

© 2003 ARTECH HOUSE, INC.
685 Canton Street
Norwood, MA 02062

All rights reserved. Printed and bound in the United States of America. No part of this book may be reproduced or utilized in any form or by any means, electronic or mechanical, including photocopying, recording, or by any information storage and retrieval system, without permission in writing from the publisher.
   All terms mentioned in this book that are known to be trademarks or service marks have been appropriately capitalized. Artech House cannot attest to the accuracy of this information. Use of a term in this book should not be regarded as affecting the validity of any trademark or service mark.

International Standard Book Number: 1-58053-160-1
A Library of Congress Catalog Card Number is available from the Library of Congress.

10 9 8 7 6 5 4 3 2 1

*To Manjur, Setara, and Farazana Chowdhury
and Marianne Sadek*

Contents

	Preface	*xv*
	Acknowledgments	*xvii*
1	**Introduction**	**1**
1.1	What Is ITS?	1
1.2	ITS Today and Tomorrow	2
1.3	ITS Training and Education Needs	2
1.4	Scope of the Book	4
1.5	Organization of the Book	5
	Review Questions	6
	Reference	6
2	**Fundamentals of Traffic Flow and Control**	**7**
2.1	Traffic Flow Elements	7
2.1.1	*Flow-Density Relationships*	8
2.1.2	*Fundamental Diagram of Traffic Flow*	9
2.2	Traffic Flow Models	11
2.2.1	*Greenshields Model*	11
2.2.2	*Alternative Traffic Flow Models*	13

2.3		Shock Waves in Traffic Streams	14
2.3.1		*Shock Wave Velocity*	15
2.3.2		*Shock Wave Analysis Example*	16
2.4		Traffic Signalization and Control Principles	18
2.4.1		*Traffic Signalization Principles*	18
2.4.2		*Actuated Signal Control*	21
2.4.3		*Signal Coordination*	22
2.5		Ramp Metering	24
2.5.1		*Types of Ramp Control*	25
2.5.2		*Ramp Metering Operational Concept*	25
2.6		Traffic Simulation	30
2.6.1		*Traffic Simulation Models*	30
2.6.2		*Examples of Traffic Simulation Models*	31
2.6.3		*Basic Guidelines for Applying Simulation Models*	32
2.7		Conclusions	32
		Review Questions	32
		References	34
3		**ITS User Services**	**35**
3.1		ITS User Services	35
3.2		ITS User Services Bundles	36
3.3		Travel and Traffic Management	36
3.3.1		*Pretrip Travel Information*	38
3.3.2		*En Route Driver Information*	38
3.3.3		*Route Guidance*	39
3.3.4		*Ride Matching and Reservation*	39
3.3.5		*Traveler Services Information*	39
3.3.6		*Traffic Control*	40
3.3.7		*Incident Management*	40
3.3.8		*Travel Demand Management*	40
3.3.9		*Emissions Testing and Mitigation*	41
3.3.10		*Highway-Rail Intersection*	41
3.4		Public Transportation Operations	42

3.4.1	*Public Transportation Management*	42
3.4.2	*En Route Transit Information*	43
3.4.3	*Personalized Public Transit*	43
3.4.4	*Public Travel Security*	44
3.5	Electronic Payment	44
3.6	Commercial Vehicles Operations	45
3.6.1	*Commercial Vehicle Electronic Clearance*	45
3.6.2	*Automated Roadside Safety Inspection*	45
3.6.3	*Onboard Safety Monitoring*	45
3.6.4	*Commercial Vehicle Administrative Processes*	46
3.6.5	*Hazardous Materials Incident Response*	47
3.6.6	*Freight Mobility*	47
3.7	Emergency Management	47
3.7.1	*Emergency Notification and Personal Security*	47
3.7.2	*Emergency Vehicle Management*	48
3.8	Advanced Vehicle Control and Safety Systems	48
3.8.1	*Longitudinal Collision Avoidance*	48
3.8.2	*Lateral Collision Avoidance*	49
3.8.3	*Intersection Collision Avoidance*	50
3.8.4	*Vision Enhancement for Collision Avoidance*	51
3.8.5	*Safety Readiness*	51
3.8.6	*Precrash Restraint Deployment*	51
3.8.7	*Automated Highway System*	52
3.9	Information Management	52
3.10	*Maintenance and Construction Management*	52
3.11	Conclusions	53
	Review Questions	53
	Reference	54
4	**ITS Applications and Their Benefits**	**55**
4.1	Freeway and Incident Management Systems	56
4.1.1	*FIMS Objectives*	56
4.1.2	*FIMS Functions*	57
4.1.3	*Traffic Surveillance and Incident Detection*	59

4.1.4	*Ramp Control*	66
4.1.5	*Incident Management*	67
4.1.6	*Information Dissemination*	71
4.1.7	*Real-World Freeway and Incident Management Systems and Their Benefits*	74
4.2	Advanced Arterial Traffic Control Systems	75
4.2.1	*Historical Development of Computer Traffic Control Systems*	75
4.2.2	*Adaptive Traffic Control Algorithms*	76
4.2.3	*Real-World Adaptive Traffic Control Systems and Their Benefits*	78
4.3	Advanced Public Transportation Systems	80
4.3.1	*AVL Systems*	80
4.3.2	*Transit Operations Software*	82
4.3.3	*Transit Information Systems*	83
4.3.4	*Electronic Fare Payment Systems*	86
4.4	Multimodal Traveler Information Systems	86
4.5	Conclusions	89
	Review Questions	89
	References	90
5	**ITS Architecture**	**93**
5.1	Regional and Project ITS Architecture	93
5.2	Why Do We Need an ITS Architecture?	94
5.3	Concept of Operations	95
5.4	National ITS Architecture	98
5.4.1	*User Services and User Service Requirements*	98
5.4.2	*Logical Architecture*	101
5.4.3	*Physical Architecture*	104
5.4.4	*Equipment Packages*	107
5.4.5	*Market Packages*	107
5.5	Proposed Procedure for Developing a Regional ITS Architecture	108
5.6	Architecture Development Tool	109

5.7	Conclusions	109
	Review Questions	111
	References	111

6	**ITS Planning**	**113**
6.1	Transportation Planning and ITS	113
6.2	Planning and the National ITS Architecture	117
6.3	Planning for ITS	119
6.3.1	*Market Package-Based ITS Planning Process*	121
6.3.2	*Traceability-Based ITS Planning Process*	121
6.4	Case Study: Northern Virginia ITS Planning Exercise Using the Traceability-Based Approach	124
6.5	Integrating ITS into Transportation Planning	125
6.6	Conclusions	132
	Review Questions	132
	References	133

7	**ITS Standards**	**135**
7.1	Introduction	135
7.2	Standard Development Process	136
7.3	National ITS Architecture and Standards	137
7.4	ITS Standards Application Areas	138
7.4.1	*Center-to-Roadside Interface Class*	138
7.4.2	*Center-to-Center (C2C) Applications*	141
7.4.3	*Center-to-Vehicle/Traveler*	144
7.4.4	*Roadside-to-Vehicle*	145
7.4.5	*Roadside-to-Roadside*	146
7.5	NTCIP	147
7.6	Standards Testing	150
7.7	Conclusions	150
	Review Questions	151
	References	151

8	**ITS Evaluation**	**153**
8.1	Introduction	153
8.2	Project Selection at the Planning Level	154
8.2.1	*Benefit-Cost Analysis*	154
8.2.2	*Relative Rating and Ranking*	155
8.3	Deployment Tracking	155
8.4	Impact Assessment	160
8.5	Benefits by ITS Components	162
8.6	Benefit Estimation Categories	162
8.7	Evaluation Guidelines	163
8.8	Evaluation Support Tools	165
8.8.1	*ITS Deployment Analysis System*	165
8.8.2	*Traffic Simulation Models*	166
8.9	Challenges for ITS Evaluation	167
8.10	Conclusions	168
	Review Questions	169
	References	169
9	**ITS Challenges and Opportunities**	**171**
9.1	Mainstreaming ITS	171
9.2	System Upgrade	173
9.3	System Integration	174
9.4	Training Needs	174
9.5	Funding	175
9.6	Privacy	175
9.7	Rulemaking and Compliance	176
9.8	Resource Sharing	176
9.9	ITS and National Security	177
9.10	Conclusions	178
	Review Questions	179

References 179

About the Authors **181**

Index **183**

Preface

The future of our transportation system looks promising with the introduction of *Intelligent Transportation Systems* (ITS). The last few years of experience with ITS deployments give us the confidence that, with widespread and more technologically advanced systems, our surface transportation will get better with time. The twenty-first century promises an exciting time for our transportation infrastructure. Our transportation system will ride on the technological advances of this century, achieved in areas such as communications, electronics and control, resulting in superior mobility and enhanced safety for the users of the surface transportation system.

ITS gives us the tools, and potential for future tools, which could make our transportation system safer. We can expect significant improvement in safety in our future transportation system. ITS would provide the path to minimize risk of crashes by integrating safety-enhancing functions within vehicles or in the infrastructure. Transportation crashes are results of infrastructure deficiency, human/vehicle operator factors, vehicle and/or environmental conditions. ITS provides the opportunity to provide functions in the vehicle and/or infrastructure to mitigate these deficiencies. For example, sensors in the mainline highway could provide advance warning about an oncoming vehicle to side street traffic on a stop-controlled intersection, to compensate for any sight distance deficiencies for the side street traffic. In-vehicle sensors could provide advance warning to inattentive or drowsy drivers before hitting another vehicle or object, or before running off the road. The possibilities to improve the safety of our transportation system are endless.

Sitting idly in traffic congestion could be a thing of the past with ITS. ITS will improve flow by managing traffic flow, detecting and clearing incidents efficiently, and providing information in an effective and efficient manner. ITS could provide priority to transit and emergency vehicles. Commercial vehicle services are improved with accurate location information, and communications between driver and operations center. Additionally, nonrisk and legal commercial vehicles could go through checkpoints without stopping through the use of electronic clearance systems.

ITS provides improvements across all modes of surface transportation. Its impacts on multimodal transportation system would provide a balanced transportation system, while benefiting all modes of surface transportation. In order to enjoy the enormous benefits in surface transportation that are promised by ITS in the future, we need careful planning at the national, state, and regional levels. ITS deployed with a carefully drawn-out plan will minimize any risk of failure, and maximize the expected benefits. Transportation professionals need to be knowledgeable in various aspects of ITS planning. Furthermore, an ITS planner should have fundamental knowledge of traffic flow theory and ITS applications.

The purpose of this book is to present fundamental knowledge on various topic areas that are necessary for successful ITS planning. The book covers diverse areas such as traffic flow and traffic control fundamentals, ITS user services and applications, regional ITS architecture, ITS planning, ITS standards, and ITS evaluation. All of these areas are critical to the success of the ITS planning process. The book is sufficiently detailed in its discussion of key aspects of ITS planning, which makes it applicable to various and diverse regions around the globe. It would also help ITS operators in understanding the major concepts of ITS applications, challenges and opportunities.

This book can serve as a main or supplementary text for upper-level undergraduate and graduate courses in ITS or related areas. Additionally, the book offers a comprehensive and easy-to-follow process that guides a reader through the fundamentals essential for ITS and operations, which allows the book to also serve as a practical reference for practitioners in transportation operations/planning or ITS operations/planning.

Acknowledgments

Several individuals provided enormous support in the preparation of the manuscript. Because of their contributions to this book, their names should be mentioned here. We would like to acknowledge the contributions of Jeff Brummond, Principal System Architect, Iteris, Inc., for providing valuable comments on various chapters of this book. He was a major inspiration in starting and completing this book. We would also like to thank Barbara Lovenvirth, Nichole Derov, Surekha Lingala, Paulin Tan, Niki Maxwell, James Pol, and Amy Tang McElwain for their support in the preparation of the manuscript.

Finally, a very special note of gratitude is expressed to our families, Manjur, Setara, and Farazana Chowdhury, and Marianne Sadek, for the support and encouragement we received during the writing of this book.

1
Introduction

This chapter is intended to introduce *Intelligent Transportation Systems* (ITS) to the reader. The chapter also discusses the training and education needs of ITS professionals and presents an overview of the chapters in this book.

1.1 What Is ITS?

ITS refers to a variety of tools, such as traffic engineering concepts, software, hardware, and communications technologies, that can be applied in an integrated fashion to the transportation system to improve its efficiency and safety. ITS provides support to improve services in transportation system operations, such as traffic management, commercial vehicle operations, transit management, and information to travelers. It provides an alternative or enhancement to traditional solutions to transportation problems. Traditionally, the transportation community attempts to meet the challenges of increasing travel demands by building additional capacity. This solution may not work in areas that have already been built up or face construction constraints due to stringent environmental regulations. In such cases, ITS can serve as a good alternative to meeting future travel demands.

ITS has a lot to offer in solving some of the nation's most complicated transportation problems. ITS has the potential to improve traffic flow by reducing congestion, improve air quality by reducing pollution and travel delay, and improve safety by providing advance warning of probable crash situations and minimizing the effects of environmental, highway, and human factors that contribute to crashes. It can also foster economic growth in a region by improving mobility and reducing fuel consumption.

In addition, ITS has great potential for making travel more convenient by providing timely and accurate information on the Internet as well as available travel options. More personalized information, such as estimated travel times and the shortest travel route to a destination, can be made available to travelers through handheld or in-vehicle devices and the Internet.

Commercial vehicle operators, regulatory and taxing agencies, and highway users can also benefit from ITS applications that support electronic administrative processes and automated roadside safety inspections. ITS applications also provide many benefits to public transit users and operators. These include improving security on transit vehicles and at transit stations, providing real-time schedule information to transit users, suggesting alternate routes for transit during an incident, and giving transit preferential treatment at traffic signals.

ITS attempts to improve the efficiency of the transportation system by using real-time and historical information on the system's status to optimally allocate resources across transportation system components. ITS applications improve the existing transportation system by allowing it to operate more safely and efficiently. In general, ITS applications have the potential to reduce travel time, reduce the frequency and severity of crashes, improve flow, reduce costs, and improve customer satisfaction. Evaluation studies and operational tests have shown that ITS applications have provided significant benefits across various surface transportation modes.

1.2 ITS Today and Tomorrow

In large metropolitan areas, ITS projects have become a part of the transportation system. ITS is currently being implemented, either independently or as part of traditional transportation projects, in many areas in order to support the objectives of a safer and more efficient transportation system. Some metropolitan areas have already seeing the benefits of ITS through improved travel conditions. As a result, many smaller metropolitan and rural areas are also serious by thinking about ITS or, in some cases, already implementing it.

ITS technology, however, is changing very rapidly. New technologies arrive every day, and the demand for a more efficient and safer transportation system continues to grow. ITS is expected to grow rapidly, both geographically and functionally, to meet future demands on the transportation system.

1.3 ITS Training and Education Needs

ITS is a new discipline. Traditional educational programs in transportation engineering at the nation's universities are not adequate to prepare students to plan, design, and operate ITS. In addition to traditional transportation engineering courses, ITS professionals require training in many diverse areas, such as systems engineering, electronics, communications systems, and institutional issues.

ITS concepts were given serious consideration in the early 1990s as a practical way to meet future demands on the transportation system. The breadth and elements of ITS are still being analyzed and developed. As ITS continues to become more and more prevalent in the transportation sector, transportation professionals are trying to identify the training and education needs to plan, design, deploy, operate, and evaluate ITS. The National ITS Professional Capacity Building Program identified 10 major critical areas where ITS training and education needs to be concentrated [1]. These areas are as follows:

- *Planning and regional concept of operations:* Planning for ITS is different from planning for traditional transportation projects, such as a highway construction. ITS planning involves a move from construction to alternative solutions incorporating advanced technologies to meet future traffic demand. The challenge to the ITS professional is to integrate ITS planning into the traditional planning process to bring ITS to mainstream transportation activities.
- *System analysis and design:* This involves the ability to identify users, define user requirements, and design a system that meets those requirements. The ability to analyze and design software and communication systems will be necessary in many projects.
- *Technology evaluation:* ITS professionals should be able to choose the most appropriate and most cost-effective strategy and technology. Furthermore, in addition to being familiar with various evaluation methods, they should also be familiar with the different technologies and their capabilities and limitations.
- *Data analysis and management:* ITS applications typically involve the collection of large amounts of data. ITS professionals should know how to analyze these data, how to extract useful information from them, and how to manage and distribute the information.
- *Systems integration:* Systems integration involves connecting individual deployments and institutions together into a comprehensive regional transportation system to optimize services provided to users. It provides maximum benefits by minimizing redundancies and maximizing capabilities through the integration of different components and institutions.
- *Organizational and institutional issues:* For ITS to succeed, ITS professionals must know about organizational and institutional issues related

to ITS deployments and the challenges they pose. Such issues include the changes required in procurement and contracting procedures for ITS in highway construction projects and coordination requirements between different agencies for an ITS project.

- *Contract management:* ITS professionals need to be trained on ITS project management approaches, which are different from those required to manage a construction project. These include knowing how to select the appropriate contractors for a project, "acceptance testing" of a project with respect to requirements, and warranty periods for new deployments. In many cases, construction contractors who may undertake an ITS project lack the background knowledge needed for successful ITS installations.

- *Financing:* ITS professionals need to know the funding sources for ITS projects. In addition, they should know how to optimize the resources of these funding sources to meet project objectives.

- *Coalition building:* Building and maintaining consensus among stakeholders is key to successful regional and statewide ITS deployments. ITS professionals should be able to engage stakeholders and develop consensus in meeting project objectives.

- *Writing/communications:* ITS professionals should be able to write specifications that will help procure the best possible system. Common problems in ITS deployment are misunderstandings and miscommunications between the deploying agency and the contractors. Many times, incomplete or ambiguous specifications lead to confusions between the deploying agency and the contractor, which can ultimately lead to project failure.

This book is designed to cover some of these areas, so as to serve as a resource in ITS training and education.

1.4 Scope of the Book

The implementation of ITS projects is quite complex. The challenge is in creating a systematic implementation plan, designing a system that is easily upgradeable and expandable, solving institutional issues, identifying and implementing operational plans, and evaluating the costs and benefits. Classical approaches to project implementation include planning, design, deployment, operations and maintenance, and evaluations. The primary focus of this book is on ITS planning and evaluations. The book is intended to serve as both an academic textbook that can help supplement an upper-level undergraduate or introductory graduate level course in ITS, as well as a reference book for practitioners. We authors have tried to provide a combination of theoretical background information along with real-world examples to serve the needs of both communities.

1.5 Organization of the Book

The book focuses on describing the planning process for ITS deployment. It describes how the process fits within the traditional transportation planning framework and discusses the tools that can be used in ITS planning. The book also includes a description of the regional ITS architecture development process, where an ITS regional architecture is developed to specify how the different ITS components would interact with one another to help solving regional transportation problems.

The book is organized in nine chapters, described as follows:

- Chapter 1 explains the meaning of the term ITS and describes the book's scope and organization.
- Chapter 2 summarizes some of the basic fundamentals of traffic flow and transportation systems operations, traffic control principles, and communications systems. These fundamental concepts are essential for appreciating ITS and how it can help improve the efficiency and safety of the transportation system.
- Chapter 3 focuses on describing the different ITS user services. The term ITS user services means what ITS can do for the users of the transportation system. These users include travelers, operators, state departments of transportation, and commercial vehicle drivers, among others.
- Chapter 4 is dedicated to giving some examples of real-world ITS applications to demonstrate that ITS is becoming a part of our national transportation system.
- Chapter 5 describes the process of developing an ITS regional architecture, which is a blueprint for deploying ITS technologies in a particular region.
- Chapter 6 discusses the process of planning for ITS deployment and describes the various steps of this process.
- Chapter 7 discusses the critical issue of ITS standards, which are needed to ensure that ITS deployments across regions are interoperable and can communicate with one another.
- Chapter 8 describes the process of ITS evaluation, which should be regarded as an integral part of ITS planning and deployment.

- Chapter 9 discusses some of ITS challenges and opportunities.

Review Questions

1. How does ITS fit within the traditional transportation system? What are the differences and synergies between traditional transportation systems and ITS?
2. What are some of the benefits of ITS?
3. How can ITS contribute to the traditional transportation system?
4. You want to hire an ITS planner for your region. Write an ad describing the potential candidate's qualifications.
5. Prepare job requirements for the following ITS professionals for a local transportation agency:
   - Designer;
   - Project manager responsible for the operation and maintenance of local ITS projects.
6. Identify the most congested corridor in your area. What are the problems associated with that corridor? What type of technology can help solve those problems?

Reference

[1] U.S. Department of Transportation, "Building Professional Capacity," *ITS: Guidelines for Designing an Individualized Training and Education Plan*, April 1999.

2
Fundamentals of Traffic Flow and Control

This chapter focuses on the fundamentals of traffic flow theory and the basic principles of traffic control and operations. A basic understanding of these fundamental concepts is imperative to fully appreciate a number of ITS applications, including freeway management systems, arterial management systems, and automated highways. This chapter is divided into six sections. Section 2.1 introduces the primary elements of a traffic stream (flow, density, and speed) and discusses flow-density relationships. Section 2.2 focuses on the mathematical relationships used for describing traffic flow. Section 2.3 discusses the phenomenon of shock waves in traffic streams and shows how the concept could be used to determine the length of a queue forming at a traffic light or as the result of an accident. Section 2.4 is devoted to the principles of traffic signalization and control. Section 2.5 focuses on freeway ramp metering, while Section 2.6 briefly discusses the use of simulation models to study complex traffic situations.

2.1 Traffic Flow Elements

Three basic elements—flow, speed, and density—are typically used to describe a traffic stream. Density is also related to the gap or headway between two vehicles in the traffic stream. A brief definition of these elements follows:

- *Flow* (q). Flow can be defined as the number of vehicles passing a given point on a highway during a given period of time, typically 1 hour (vehicles per hour). An important flow parameter is the maximum flow value, which is often referred to as the capacity (q_m) of a roadway section.

- *Speed* (u). Speed is the distance traveled by a vehicle during a unit of time. Speed is usually expressed in miles per hour (in the United States), kilometers per hour, or feet per second. There are two important speed parameters: the free-flow speed (u_f) and the optimum speed (u_o). The free-flow speed (u_f) is the absolute maximum speed that is attained when the flow approaches zero (i.e., only one vehicle exists on the highway). The optimum speed, on the other hand, is the speed of the traffic stream under maximum flow conditions (i.e., capacity conditions).

- *Density* (k). Traffic density can be defined as the number of vehicles present over a unit length of a highway at a given instant in time. Density is typically expressed in vehicles per mile (in the United States) or vehicles per kilometer. There are two important density parameters: the jam density (k_j) and the optimum density (k_o). The jam density (k_j) occurs under extreme congestion conditions when the flow and speed of the traffic stream approach zero. The optimum density (k_o) occurs under maximum flow conditions.

- *Headway.* Headway can generally be defined as the time or distance gap between two vehicles in the traffic stream. The time headway (h) is defined as the difference in time between the moment the front of a vehicle arrives at a point on the highway and the moment the front of the following vehicle arrives at that same point. The time headway is typically expressed in seconds. The space headway (d), on the other hand, is defined as the distance between the front of a vehicle and the front of the following vehicle (in feet or meters).

2.1.1 Flow-Density Relationships

The three basic parameters of a traffic stream (flow, speed, and density) are related to each other by the following equation:

$$\text{Flow} = \text{speed} \times \text{density}$$
$$q = uk \qquad (2.1)$$

This equation states that the flow is equal to the product of speed and density. So, for example, if a 1-mile stretch of roadway contains 10 vehicles (i.e., $k = 10$), and the mean speed of the 10 vehicles is 50 mph, after 1 hour, 500 vehicles (50 × 10) would have passed. In other words, the value of the flow (q) in this case equals 500 *vehicles per hour* (vph).

2.1.2 Fundamental Diagram of Traffic Flow

The relationship between density and flow is generally referred to as the fundamental diagram. The following hypotheses could be made regarding this relationship:

1. At a value for the density equal to 0 (i.e., no vehicles exist on the highway), the flow is also going to be equal to 0.
2. As the density increases, the flow also increases.
3. When the density reaches its maximum value (i.e., the jam density, k_j), the flow must be zero.
4. It thus follows from (2) and (3) that as the density increases, the flow initially increases up to a maximum value (q_m). Further increases in density will lead to a reduction of the flow, which will eventually become zero when the density is equal to the jam density.

The shape of the relationship between flow and density takes the general form shown in Figure 2.1(a). Since, from (2.1), the speed (u) could be expressed as the flow/density (q/k), it follows that speeds at a given point on Figure 2.1(a) could be represented by radial lines from the origin to that point, as the figure shows.

Similar hypotheses can be made regarding the relationship between speed and density and the relationship between speed and flow. For the speed-density relationship, when the density approaches zero (i.e., there is little interaction between individual vehicles), drivers are free to select whatever speed they desire,

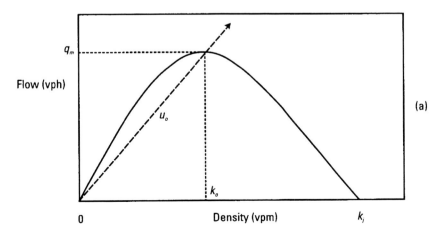

Figure 2.1 (a) Relationship between flow and density; (b) relationship between speed and density and; (c) relationship between speed and flow.

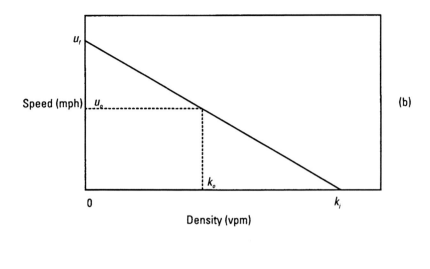

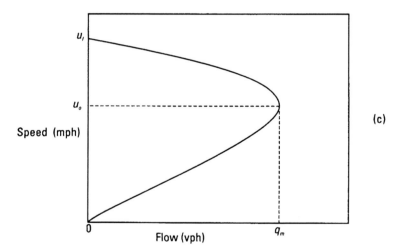

Figure 2.1 Continued

and hence the corresponding speed is equal to the free-flow speed (u_f). As the density increases, the speed decreases until it reaches a value of zero when the road is completely jammed [i.e., when the density is equal to the jam density (k_j)]. Figure 2.1(b) depicts this general relationship between speed and density.

Similarly, for the relationship between speed and flow, one could assume that the speed would be equal to the free-flow speed (u_f) when the density and flow equals zero. Continuous increase in flow will then result in a continuous decrease in speed. There will be a point, however, when the further addition of vehicles will result in a reduction in the number of vehicles passing a given point on the highway (i.e., reduction in the flow). The addition of vehicles beyond this point would result in congestion, and both the flow and speed would decrease until they both become zero. The relationship between speed and flow could then be represented as shown in Figure 2.1(c).

Figure 2.1 shows that it would be desirable to operate roadways at densities not exceeding the density at capacity, in order to avoid congestion. Section 2.2 presents some mathematical models that can be used to describe the relationships between the traffic flow elements, as depicted in Figure 2.1.

2.2 Traffic Flow Models

Generally speaking, mathematical models describing traffic flow could be divided into macroscopic flow models and microscopic flow models. Macroscopic models are concerned with describing the flow-density relationship for a traffic stream (a group of vehicles). Microscopic models, on the other hand, describe flow by tracking individual vehicles using car-following logic. The primary focus in this chapter is on macroscopic models.

2.2.1 Greenshields Model

Greenshields was among the first researchers attempting to develop models for describing traffic flow. Greenshields postulated that a linear relationship exists between speed and density having the following form:

$$u = u_f - \frac{u_f}{k_j} k, \qquad (2.2)$$

where all the terms are as defined above. As previously discussed, this equation indicates that as the density (k) approaches zero, the speed (u) approaches free-flow speed (u_f). Also, as the speed (u) approaches zero, the density approaches jam density (k_j).

From (2.1), we know that $q = uk$. Therefore, using the Greenshields equation (2.2), the flow (q) could be expressed as

$$q = u_f k - \frac{u_f}{k_j} k^2 \qquad (2.3)$$

Also, from (2.1), we know that $k = q/u$. Therefore, substituting q/u for k, the relationship between the speed (u) and the flow (q) could be expressed as

$$u^2 = u_f u - \frac{u_f}{k_j} q \qquad (2.4)$$

Equations (2.2) to (2.4) describe the three diagrams depicted in Figure 2.1(a–c), respectively. As can easily be seen, the three equations or diagrams are rather redundant, because if only one relation is known, the other two could very easily be derived using the basic relation that states that $q = uk$, as previously described. Nevertheless, each of the three diagrams has its own use. For theoretical work, the relationship between speed and density is the one typically used, since there is only one value for the speed for each value of the density. This is not the case with the other two diagrams. The relationship between flow and density is used in freeway and arterial control systems to control the density in an effort to optimize productivity (flow). Finally, the relationship between speed and flow could be used for design to define the trade-off between the level of service on a road facility (as expressed by the speed), and the productivity (as defined by the flow).

Determining the Density and Speed at Capacity

Equations (2.3) and (2.4) could be used to determine the density and speed at maximum flow conditions (k_o and u_o, respectively). To determine the density at maximum flow, (2.3) is differentiated with respect to the density, and the resulting expression is equated to zero, as follows:

$$\frac{dq}{dk} = u_f - 2k \cdot \frac{u_f}{k_j}$$

Therefore,

$$0 = u_f - 2k_o \cdot \frac{u_f}{k_j}$$

or,

$$k_o = \frac{k_j}{2} \qquad (2.5)$$

This means that, based on the Greenshields model, the density at capacity is equal to one half of the jam density.

Similarly, to determine the speed at capacity, we would differentiate (2.4) with respect to u, and equate dq/du to zero, as follows:

$$2u = u_f - \frac{u_f}{k_j}\frac{dq}{du}$$

Therefore,

$$\frac{dq}{du} = k_j - 2u_o\frac{k_j}{u_f} = 0$$

or,

$$u_o = \frac{u_f}{2} \tag{2.6}$$

This means that, for a Greenshields model, the speed at capacity is equal to one half the free-flow speed. From (2.5) and (2.6), and since the flow is equal to the product of the speed and density, the maximum flow (i.e., capacity) could be expressed as

$$q_m = \frac{k_j u_f}{4} \tag{2.7}$$

Once again, it should be noted that this result is only valid for the Greenshields model.

2.2.2 Alternative Traffic Flow Models

The Greenshields model is not the only model available for modeling traffic flow. Over the years, several researchers proposed different forms of mathematical models to describe traffic flow. Earlier models, similar to the Greenshields model, assumed a single regime phenomenon over the entire range of traffic flow conditions (i.e., they used the same equation to model both noncongested or free-flow conditions and congested conditions). More recent models attempted to refine earlier models by considering two separate regimes for the free-flow and congested-flow regimes. Examples of single-regime models include the Greenshields model, the Greenberg model, the Underwood model, and the Northwestern model. Multiregime models, on the other hand, include Edie's model, the two-regime linear model, the modified Greenberg model, and the three-regime linear model. A discussion of these models is beyond the scope of this introductory chapter on traffic flow fundamentals. Interested readers can refer to [1] for more details.

In addition to macroscopic models to describe traffic flow, other models were also developed to describe how one vehicle would follow another vehicle. These models are typically called *car-following models* or *microscopic flow models* (since they deal with individual vehicles and not with a traffic stream). Researchers have shown that macroscopic and microscopic models are related. For a discussion of this relationship, readers can also refer to [1].

2.3 Shock Waves in Traffic Streams

Traffic conditions (as defined by the flow, speed, and density) change over time and space. When these conditions change, a boundary is established that demarks or distinguishes one flow state from another. In the context of traffic flow theory, this boundary is called a *shock wave*. Consider, for example, the case of a single-lane approach to a pretimed signalized intersection (Figure 2.2). During the red interval, just upstream the stop-line, vehicles will be stopped and a queue will form. However, at some distance upstream of the signal, original traffic conditions will prevail, and immediately downstream of the signal, free-flow conditions will exist. This means that in this simple example, we have three distinct traffic states: (1) original traffic flow conditions at some distance upstream of the signal (condition A); (2) platooning conditions within the queue forming at the signal (condition B); and (3) free-flow conditions downstream the signal (condition C).

These three distinct conditions will give rise to two shock waves. The first shock wave (wave AB) establishes the boundary between original flow conditions upstream the signal (condition A), and conditions within the platoon (condition B). This shock wave defines the rear end of the queue forming at the stop line and is moving upstream, in a direction opposite to traffic. Shock wave AB is referred to as a backward forming shock wave (it is labeled forming because its propagation results in the formation of a queue). The second shock wave, wave BC, marks the boundary between conditions within the platoon (condition B) and free-flow conditions downstream the signal (condition C).

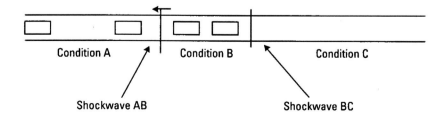

Figure 2.2 Shock waves at a signalized intersection during the red interval.

This second shock wave is stationary and defines the front end of the queue. The rate at which the forming queue grows is related to the relative velocity between the front shock wave (AB) and the rear shock wave (BC). Since shock wave BC is stationary, in this case, the queue will grow at a rate equal to the velocity of shock wave AB.

Now, when the signal turns green, vehicles will start discharging from the front end of the queue, resulting in a new traffic state downstream from the signal that we refer to in Figure 2.3 as the release condition (condition D). A new shock wave (shock wave BD) will form in this case, marking the boundary between platoon conditions (condition B) and the release condition (condition D). Shock wave BD will also move in a direction opposite to traffic, but in this case, the shock wave is actually helping decrease congestion. Given this, shock wave BD is typically referred to as a backward recovery shock wave. The queue will totally dissipate when the backward recovery shock wave (BD) catches up with the backward forming shock wave (AB).

2.3.1 Shock Wave Velocity

Consider the section of the highway shown in Figure 2.4 with two different traffic states A and B. The figure shows three different speeds: the mean speed of traffic stream A, u_A; the mean speed of traffic stream B, u_B; and the speed of the shock wave separating the two traffic conditions, u_w. At the shock wave boundary, the number of vehicles leaving flow condition A should be equal to the number of vehicles entering flow condition B, since vehicles are neither created nor destroyed.

The number of vehicles leaving state A (N_A) can be given as

$$N_A = (u_A - u_w) \cdot k_A \cdot t$$

where $(u_A - u_w)$ is the velocity of vehicles in flow condition A relative to the shock wave velocity, and t is a given time interval.

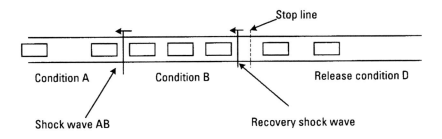

Figure 2.3 Shock waves at signalized intersections when light turns green.

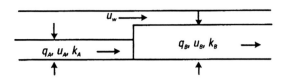

Figure 2.4 Shock wave analysis. (*After:* [2].)

Similarly, the number of vehicles entering state B (N_B) can be expressed as:

$$N_B = (u_B - u_w) \cdot k_B \cdot t$$

where $(u_B - u_w)$ is the velocity of vehicles in flow condition B relative to the shock wave velocity.

Equating N_A to N_B gives:

$$(u_A - u_w) \cdot k_A \cdot t = (u_B - u_w) \cdot k_B \cdot t$$

or,

$$u_w = \frac{(q_B - q_A)}{(k_B - k_A)} \tag{2.8}$$

This means that the shock wave speed between two speeds is equal to the change in flow between the two states divided by the change in density. It is noted that since (2.8) was derived assuming a forward moving shock wave, it should be possible to use the sign of u_w derived from (2.8) to determine the direction of the shock wave propagation. If u_w is positive, this means that the shock wave is moving in the direction of traffic. If it is negative, the wave is moving opposite the direction of traffic.

Shock wave analysis could be used to analyze many traffic situations. For example, we could use shock wave analysis to determine the length of a queue forming at a work zone or as a result of an incident on a freeway, as well as to determine the time it takes for the formed queue to dissipate. Section 2.3.2 illustrates how the analysis can be conducted.

2.3.2 Shock Wave Analysis Example

A vehicular stream (*A*) has the following characteristics: a flow value (q_a) equal to 1,200 vph and a density (k_a) equal to 100 vpm. The vehicular stream is interrupted by a flag person for 5 minutes. After 5 minutes, vehicles at the front of the stationary platoon begin to be released at $q_b = 1,600$ vph and $u_b = 20$ mph. Assuming that the jam density (k_j) is equal to 240 vpm, determine (1) the length of the queue forming at the end of the 5 minutes; and (2) the time it takes for the queue to dissipate.

Determine the Length of the Queue Form

With reference to Figure 2.5, in order to determine the length of the queue, we first need to determine the velocity of shock wave AQ defining the rear end of the queue. Shock wave AQ marks the boundary between flow condition A ($q = 1,200$ vph, and $k = 100$ vpm), and conditions within the stopped queue ($q = 0$ vph, and $k = k_j = 240$ vpm).

Therefore, using (2.8),

$$U_{QA} = (0 - 1,200)/(240 - 100) = -8.57 \text{ mph}$$

Since the front shock wave defining the front end of the queue is stationary, the queue will grow at a rate equal to the speed of the rear shock wave (u_{AQ}). The length of the queue at the end of the 5 minutes is given by:

$$8.57 \times 5/60 = 0.714 \text{ mile}$$

After the flag person moves away, a recovery shock wave develops as vehicles discharge from the queue, as shown in Figure 2.6. The recovery shock, QB,

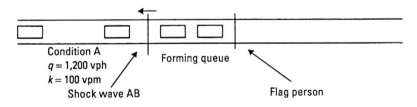

Figure 2.5 Queue formation.

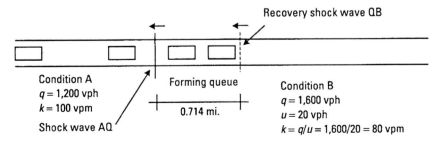

Figure 2.6 Recovery shock wave.

marks the boundary between queuing conditions within the queue ($q = 0$ and $k = 240$ vpm), and release condition B ($q = 1,600$ vph, $u = 20$ vph, and $k = q/u = 1,600/20 = 80$ vpm).

The speed of shock wave, QB, is thus given by

$$U_{QB} - (1,600 - 0) / (80 - 240) = -10 \text{ mph}$$

The queue will dissipate when the recovery shock wave, QB, catches up with shock wave AQ. Therefore, the time needed for the queue to dissipate is given by:

$$\text{Time} = 0.714 / (10 - 8.57) = 0.50 \text{ hour or 30 minutes}$$

2.4 Traffic Signalization and Control Principles

At-grade intersections represent one of the most complex components of a transportation system. This is because conflicting traffic streams compete for the right of way at an intersection. When traffic volumes are low, traffic at an intersection could be regulated using basic rules-of-the-road (e.g., one common rule states that the driver on the left must yield to the driver on the right), or using STOP and YIELD signs. As traffic volumes increase, however, it becomes extremely difficult for drivers to select adequate gaps in conflicting traffic streams to execute their desired maneuvers. When this happens, intersection signalization (i.e., the use of traffic control signals) becomes a must.

Traffic signals play a dramatic role in determining the overall performance level of an arterial system. Poorly designed traffic signals can result in unnecessary and excessive delays. On the other hand, if appropriately designed, a traffic control signal could provide for the orderly movement of traffic and could actually increase the traffic-handling capacity of an intersection. This section provides some background information on the principles of traffic signalization and timing.

2.4.1 Traffic Signalization Principles

Traffic signals could generally be divided into two groups: pretimed and actuated signals. Pretimed signals are typically insensitive to current volumes. In other words, the cycle length of a pretimed signal is generally fixed, regardless of current volumes. The operation of actuated controllers, on the other hand, varies according to the observed volume. Actuated controllers need to be connected to traffic detectors (e.g., loop detectors embedded in the pavement) to sense traffic demand and adjust signal timing accordingly.

Signals at intersections could also be divided into "isolated" intersections and intersections that operate as part of a coordinated system of signals. In general, an intersection is considered "isolated" if the distance from that intersection to nearby intersections is greater than half a mile (approximately 800m). With coordinated systems, groups of signals are timed so that moving vehicles would not have to stop at each and every intersection. Traffic signal systems have their own jargon. Some of the most important terms commonly used in this area are defined as follows:

- *Cycle and cycle length.* A traffic signal cycle is a rotation through all the signal indications at a given intersection. Every legal movement would generally receive the "green" indication only once during a given cycle. The time it takes for the signal to go through one cycle of indications is the cycle length.
- *Interval.* An interval is a time period during which all indications or lights remain unchanged. A cycle generally includes several indications such as the green interval, the change or yellow interval, the clearance or the all-red interval, and the red interval.
- *Phase.* A phase is a set of indications (i.e., the green and yellow intervals) during which a given set of movements is assigned the right of way.
- *Offset.* The offset is a term used in conjunction with coordinated systems. It refers to the time difference between the initiation of green on two adjacent signals. Typically, the offset is measured in terms of the green initiation time of the downstream (t_d) signal relative to the upstream signal (t_u); that is to say, the offset is equal to $t_d - t_u$.

2.4.1.1 Modes of Operation

Traffic signals are capable of operating in a number of different modes. The best operation mode for a given intersection is a function of several parameters, including the location, configuration, and traffic conditions at that particular intersection. The different modes of operation for a traffic signal are described as follows:

- *Pretimed operation.* Under this mode, the signal cycle length, intervals, and phases are predefined or fixed and are therefore insensitive to current traffic volumes. A pretimed signal, however, could have different "pretimed" plans for different times of the day (typically three plans for the morning peak, evening peak, and off-peak periods).
- *Semiactuated operation.* This mode of operation is used at intersections where a "major" and a "minor" street could be clearly identified. In

such cases, traffic detectors are used only on the approaches of the "minor" road, and the "major" road maintains the green until vehicles are detected on the minor street. When vehicles detected on the minor approaches, and provided that the "minimum green time" for the major road has been satisfied, the green is transferred to the minor street. As can be seen, semiactuated signals are likely to have widely varying cycle lengths depending on the arrival pattern of vehicles on the minor street approaches.

- *Full-actuated operation.* This operation mode requires detectors on all approaches. With full detection, the green is allocated based on observed volumes on a cycle-to-cycle basis. Full-actuated signals could alter the cycle length, sequence, and duration of all intervals and phases.
- *Computer-based control.* This mode refers to the use of a computer to link the operation of a group of signalized intersections into a coordinated system. The computer selects or computes "optimal" coordinated signal plans for the whole system based on the traffic information provided by the system traffic detectors. Individual signals within that system typically operate on a pretimed basis. However, they could also operate on a "quasiactuated" basis that would alter the allocation of the green among the competing volumes, while keeping the cycle length fixed.

2.4.1.2 Signal Timing Principles

On a very basic level, developing signal timing plans is based on two concepts: the "time budget" and the "critical lane." The time budget concept is concerned with allocating the available time among competing vehicular and pedestrians streams at an intersection. The critical lane concept states that during any given phase, while several traffic approaches are allowed to move, one particular movement will require the largest amount of time. That particular movement is referred to as the critical lane for that given phase. Satisfying the needs of the critical lane movement would automatically satisfy the needs of all other accompanying movements.

Typically when designing a signal plan, a traffic engineer attempts to meet the needs of the critical lane movement for each phase, while maximizing the effectiveness of the intersection. For isolated intersections, delay is commonly the measure of effectiveness used to characterize how well the intersection is performing. Coordinated systems, on the other hand, typically try to minimize a "penalty" function that represents a weighted combination of the number of stops and the total delay.

The details of developing optimal signal plans for isolated and coordinated signals are outside the scope of this book. Interested readers can refer to any standard traffic engineering textbooks [3–5] for more detailed discussions of the subject. It should also be noted that a number of computer programs are currently available to aid with this process. For the design and analysis of isolated intersections, the *highway capacity software* (HCS), which automates the procedures included within the *highway capacity manual* (HCM), is the most frequently used tool. For coordinated systems, some of the most widely used computer tools are TRANSYT-7F and SYNCRO, which will be discussed later in this section.

2.4.2 Actuated Signal Control

As previously discussed, the idea behind the use of actuated controllers is to have an adaptive type of control that is responsive to continuously changing traffic conditions. For pretimed controllers, the implemented signal plan is only optimal for the volumes assumed in developing the "off-line" plan. These assumed volumes could be very different from the actual volumes, particularly if signal plans are not updated regularly, which, unfortunately, is often the case. Actuated controllers, on the other hand, are capable of optimizing the allocation of time based on real traffic volumes.

For each actuated phase of a traffic signal, a number of parameters need to be set, among the most important of which are the following:

- *Minimum green.* Each signal phase is assigned a minimum green time. This time is typically taken to be equal to the time it takes a queue of vehicles potentially stored between the stop line and the approach detector location to enter the intersection.

- *Passage time interval.* The passage time interval is the time it takes a vehicle to travel from the detector location to the stop line. The passage time also defines the maximum gap, which is the maximum time period allowed between vehicles' arrivals at the detector for the approach to retain the green. If a time period equal to the passage time interval elapses without vehicle actuations at the detector, the green for that approach is terminated, and another approach, with a call for service waiting, gets the green. In such a case, the terminating phase is said to have "gapped out."

- *Maximum green time.* In addition to assigning each phase a minimum green, each phase is assigned a maximum green. If the demand on one approach is sufficient to retain the green until this limit (i.e., vehicles keep arriving before the maximum gap expires), the phase is terminated after the maximum green time is exceeded. In this case, the terminating phase is said to have "maxed out."

Figure 2.7 shows the basic operational concept of an actuated controller. When a certain phase becomes active, the minimum green is displayed first. Following this, the green is extended by the vehicle passage time. Depending on vehicle actuations, the minimum green is extended by the passage time interval for each vehicle actuation. If a subsequent actuation occurs within one passage time interval, another passage time interval is added (measured from the time of the new actuation and not from the end of the previous passage time interval). Finally, the green is terminated according to one of two mechanisms: (1) a passage time elapses without a vehicle actuation (the phase gaps out); or (2) the maximum green time for that phase is exceeded (the phase gaps out).

A key factor that plays a major role in defining the values of the aforementioned design parameters of an actuated controller is the location of the detector or, more specifically, the distance between the detector and the stop line. Once this distance is fixed, the different design parameters could be easily determined. For details, interested readers can refer to [4].

2.4.3 Signal Coordination

Signal coordination refers to coordinating the green of a series of traffic signals along a corridor to allow for the efficient movement of vehicles with the minimal number of stops. When signals are relatively close to one another (in

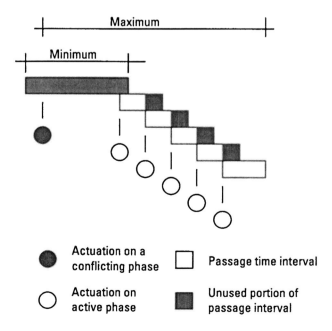

Figure 2.7 Actuated control operational concept. (*After:* [6].)

general, less than half a mile apart), their green times should be coordinated to increase traffic flow efficiency. Generally speaking, signals operating as a part of a coordinated system should all have the same cycle length. Signal coordination, when appropriately designed, can lead to significant improvement in traffic flow conditions, especially in terms of the number of stops and total delay. It can also result in a reduction in air pollution and improved air quality. However, there are several factors that limit the benefits of signal coordination. These include very short distances between intersections, complicated intersection phase plans, and a significant percentage of turning vehicles.

Given the fact that coordination requires all signals to have a common cycle, coordinated signals typically operate on a pretimed basis. This does not mean, however, that actuated controllers cannot be part of a coordinated system. In fact, they are often used within a coordinated system to alter the allocation of the green within a fixed cycle length.

A number of computer programs are currently available to help in the design of optimal signal timing plans for coordinated systems. The basic idea behind these programs is to find the set of timing parameters (such as the offsets, cycle lengths, and phase intervals) that would minimize a given performance measure (such as total delay or total number of stops), while satisfying the different constraints. Among the most famous of these computer programs are TRANSYT-7F and SYNCHRO. A brief discussion of these two programs follows.

2.4.3.1 TRANSYT-7F

TRANSYT (TRAffic Network StudY Tool) was first developed by the United Kingdom Transport Road Research Laboratories in the late 1960s and has undergone several revisions since then [7]. TRANSYT, version 7 was Americanized for the *Federal Highway Administration* (FHWA) in the late 1970s and early 1980s; hence the name, TRANSYT-7F. Currently, TRANSYT-7F is one of the most widely used computer programs for developing optimal timing plans for corridors and networks.

The TRANSYT model has two major functions: (1) the simulation of traffic flow, and (2) the optimization of traffic signal plans. For simulation, TRANSYT employs a macroscopic simulation model that models platoons of vehicles rather than individual vehicles. This model uses a platoon dispersion algorithm that simulates the dispersion or the spreading out of vehicles as they travel along road segments. Using this model, TRANSYT was shown to be able to very realistically model traffic flow on arterials and in road networks and to estimate a number of important traffic flow performance measures such as delay, stops and queuing, fuel consumption, and travel time.

The second function of TRANSYT is signal plan optimization. To develop such optimal plans, TRANSYT varies the cycle length, the phase lengths, and the signal offsets until the plan that optimizes a user-defined objective function is identified. A number of objective functions in TRANSYT include the following:

1. Minimizing a disutility index defined as a combination of stops and delay;
2. Minimizing the disutility index defined above with the inclusion of a penalty assigned for queue spillback conditions;
3. Maximizing progression opportunities along a defined corridor;
4. Maximizing the throughput (i.e., the sum of vehicles leaving the stop line).

With the objective function defined, TRANSYT uses a hill climbing search algorithm to identify the "optimal" signal plan.

2.4.3.2 SYNCHRO

SYNCHRO is another signal timing software that could be used to generate optimal signal plans (cycle length, phase lengths, and offsets). For optimization, SYNCHRO uses an objective function that attempts to minimize a combination of delay, number of stops, and number of vehicles in queue. A unique advantage of SYNCHRO is its ability to accurately model the operation of actuated controllers within a coordinated system.

2.5 Ramp Metering

Ramp metering is a form of traffic control that is applied to freeway systems. It involves regulating vehicles' entry onto the freeway system through the use of traffic signals, signs, and gates. Ramp metering's objectives include the following:

1. To balance demand and capacity and to minimize operational breakdowns;
2. To improve safety when some geometric deficiencies exist.

Ramp control systems are currently operating in many areas around the country, including Minneapolis and St. Paul, Seattle, and Austin, Texas. Most of these systems have achieved their goals in terms of reducing delay and improving safety. This section focuses on the different types of ramp control and its operational concept.

2.5.1 Types of Ramp Control

Three different types of ramp control can be identified in the literature. These include entrance ramp metering, entrance ramp closure, and exit ramp closure. Entrance ramp metering, however, is by far the most popular form of ramp control. A brief description of these three types follows.

2.5.1.1 Entrance Ramp Metering

Entrance ramp metering involves determining a metering rate according to which vehicles will be allowed entry to the freeway system. For single-lane ramps, the metering rate typically ranges between 240 and 900 vph. Ramp metering rates could be fixed for certain time periods; this is commonly referred to as *pretimed ramp metering*. In pretimed ramp metering, rates are determined based on historical observations. Rates could also vary depending on detected traffic demand (traffic responsive operations). Finally, several ramps could be controlled together as a part of an integrated system, whereby metering rates would be determined based on traffic measurements over a large segment of the freeway (systemwide ramp control).

2.5.1.2 Entrance Ramp Closure

Entrance ramp closure involves the temporary closure of the ramp using automatic gates or manual barriers. Obviously, this form of control is extreme and is likely to face strong opposition from the public. Given this, it is not very popular and should only be applied during extreme conditions.

2.5.1.3 Exit Ramp Closure

This third form of control involves closing the exit ramp using automatic gates. This form of control could be used when, for example, the cross street has inadequate capacity and the control agency would like to encourage drivers to exit elsewhere where capacity is available. Similar to entrance ramp closure, this form of control is also quite restrictive and not very popular. The rest of this section will focus on entrance ramp metering.

2.5.2 Ramp Metering Operational Concept

As previously mentioned, ramp metering is intended to limit the rate at which vehicles can enter a freeway system. Three different types of ramp metering can be identified: (1) pretimed metering; (2) traffic-responsive metering; and (3) systemwide metering. These types are described as follows.

2.5.2.1 Pretimed Metering

In pretimed metering, metering rates are predetermined based on historical observations and are not influenced by current mainline volumes. Typically, however, different rates would be established for different times of day. The calculation of the metering rate would depend on the main objective of ramp metering; that is to say, whether metering is designed to reduce congestion or improve safety.

If the system is intended for relieving congestion, the rates would be determined in such a fashion that would ensure that the demand is less than the capacity. In other words, the metering rate will be a function of the upstream demand, the ramp volume desiring to enter the freeway, and the downstream capacity. Obviously, a number of other factors will have to be taken into consideration. For example, one would have to ensure that there is adequate storage available on the ramp to accommodate the queue of vehicles forming. In addition, one would need to ensure that adequate additional capacity is available elsewhere in the corridor to accommodate any diverted demand.

On the other hand, if the system were intended to improve safety, the metering rate would be determined by the merging conditions at the ramp. At ramps, there is always a greater likelihood for rear-end and lane-change collisions as a result of vehicles' platoons on the ramp attempting to merge with main line traffic. Ramp metering can help alleviate this problem by breaking down the platoon. The appropriate metering rate in this case would depend on the geometrics of the ramp and the availability of acceptable gaps in the freeway traffic stream.

System Layout

The typical components of a pretimed ramp metering system are shown in Figure 2.8. As can be seen, the system consists of the following:

1. A ramp metering signal, which could take the form of a standard three-section (i.e., red-yellow-green) signal display or a two-section display (red-green);
2. Local controller, which is typically similar to controllers used at signalized intersections;

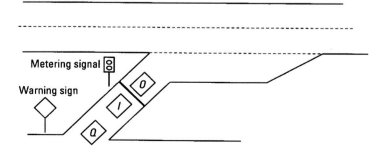

Figure 2.8 Pretimed ramp meter. (*After:* [8].)

3. Advance ramp control warning sign to inform drivers that the ramp is being metered;
4. Vehicle detectors, which are of four types: (1) check-in detectors; (2) checkout detectors; (3) queue detectors; and (4) merge detectors. These four types are described as follows:

   - *Check-in detectors (I).* With a check-in detector in use, the ramp signal would remain red until a vehicle is detected. However, a minimum metering rate (e.g., 3 vpm) would typically be imposed to avoid problems associated with possible detector failure or with a vehicle stopping too far from the stop line to actuate the detector.
   - *Checkout detectors (O).* These detectors are used to ensure single-vehicle entry. When a vehicle is allowed to pass the ramp, it is detected by the checkout detector. When this happens, the green is terminated. This ensures that the green interval is enough for the passage of one vehicle only.
   - *Queue detectors (Q).* Queue detectors are used to detect any backing or spillback of the ramp traffic onto the surface roads. When a queue is sensed for a specified period of time, the ramp metering rate may be increased to allow for shortening the queue.
   - *Merge detectors.* Merge detectors are sometimes used to detect the presence of vehicles in the merge area. When a vehicle is sensed blocking the merge area, the ramp signal may be held red until the detected vehicle manages to merge with freeway traffic. Merge detectors can help reduce the possibility of rear-end collisions.

System Operation

Under pretimed ramp control, the ramp signal would operate according to a predefined plan for the period under consideration. The timing of the red, yellow, and green intervals would depend, however, on whether the type of metering used is single-entry, platoon metering, or two-abreast metering.

- *Single-entry.* In single-entry metering, only one vehicle is allowed per green interval. The green interval (or the green-plus-yellow) is typically in the order of 1.5 to 2.0 seconds. The red interval duration will then depend on the metering rate in effect. For example, for a metering rate of 600 vph or 10 vpm, the length of the red interval would be equal to 4.0 seconds.

- *Platoon metering.* Platoon metering is used when metering rates greater than 900 vph are required. In these cases, two or more vehicles will be allowed to enter the freeway per each green interval, and the length of the green interval will have to be adequate to accommodate that number of vehicles. For example, for a metering rate of 16 vpm, and assuming that we will allow for the release of two vehicles per cycle, we will need eight cycles per minute, and the cycle length would be equal to 7.5 seconds.

- *Two-abreast metering.* In this case, two vehicles are released side-by-side, which requires two lanes on the entrance ramp. Vehicles are released alternately, and the green interval length would be set so as to allow for the release of only one vehicle per cycle. With two-abreast metering, metering rates as high as 1,700 vph can be achieved.

Pretimed ramp metering is less complex and less costly compared to traffic-responsive metering. However, it is incapable of responding to significant changes in demand. It is also incapable of responding automatically to changing traffic conditions resulting from incidents.

2.5.2.2 Traffic Responsive Metering

In traffic responsive metering, metering rates are not prefixed; instead, they are selected in real time, based on real-time traffic measurements. Traffic responsive metering systems typically use the traffic flow models previously discussed in this chapter. These models give the relationship between flow rate (q), speed (u), and density (k).

Basic Strategy

The purpose of ramp metering, in general, is to prevent congestion or to keep the values of the traffic flow parameters (i.e., flow, speed, and density) within the uncongested portion of the traffic flow parameter. Given this, the basic strategy of traffic responsive metering could be summarized as follows:

1. To obtain real-time measurements of current traffic flow parameters;
2. To determine the current state of traffic flow from the fundamental traffic flow diagram;
3. To determine the maximum ramp metering rate that would ensure that the flow is kept within the uncongested portion of the diagram.

Ramp metering strategies differ from one another based on which traffic flow parameters they use in order to determine the appropriate ramp-metering rate. Two of the most widely used traffic responsive ramp metering strategies are demand-capacity control and occupancy control.

- *Demand-capacity control.* Under demand-capacity control, ramping rates are based on real-time comparisons of upstream traffic volumes against downstream capacity. The upstream volume is measured in real-time, the downstream capacity is determined either on the basis of historical data or computed in real time based on downstream volume measurements. The ramp metering rate for the next control period (typically 1 minute) is computed as the difference between the downstream capacity and the upstream volume to ensure that downstream capacity is not exceeded. If the upstream volume exceeds the downstream capacity, the minimum ramp metering rate is implemented. One problem with using volume alone as the traffic flow performance measure is the fact that low volume values could be associated with free-flow conditions as well as with congested conditions, depending on whether traffic density is less than or greater than the density at capacity. To overcome this problem, occupancy measurements are typically taken to distinguish between congested and uncongested conditions.
- *Occupancy control.* In occupancy control, metering rates are selected based on occupancy measurements upstream the metered ramp. One out of several predefined metering rates is then selected for the next control period to suit the observed occupancy. The predefined metering rates are determined from a study of the relationship between volumes and occupancy for the facility of interest.

System Layout

Traffic-responsive ramp metering systems contain essentially the same components of pretimed metering systems (see Figure 2.7). However, the controller unit will have to have more advanced logic to monitor the traffic flow parameters and to determine the appropriate metering rates. In addition, some traffic-responsive ramp metering systems include detectors for determining traffic mix and weather conditions; this allows such systems to take the impact of these factors on traffic flow into account when determining metering rates.

2.5.2.3 System Metering

System traffic-responsive metering is the application of traffic responsive metering strategies to a series of ramps (hence the name system metering). For each control interval, real time traffic measurements of traffic parameters such as volume or occupancy are taken and used to define the demand-capacity conditions at each ramp. Using these measurements, ramp metering rates are determined for both the whole system, as well as for the individual ramp meters. The more restrictive rate, out of the local and systemwide rates, is then adopted for the next control interval.

Computing the local and systemwide metering rates follows a similar procedure to that used for pretimed ramp meters. The difference, however, is that in this case computations are conducted in real time, or based on matching predefined metering rates to real-time traffic measurements. Calculations are typically based on an optimization model such as linear programming.

System Layout

The feature that distinguishes system metering from individual pretimed or traffic-responsive metering is the interconnection among the local ramp controllers, which allows for the conditions at a given ramp to affect the rate at another location. With the controllers interconnected, the metering plans are computed by a central controller, which then communicates the developed plan to local controllers.

2.6 Traffic Simulation

Simulation is a tool for studying complex systems in the laboratory rather than in the field. In simulation, the goal is to build a model that approximates the behavior of the real system and then experiment with the model in an attempt to infer the properties concerning the behavior of the actual system. Models could be physical or mathematical; however, our focus in this chapter is on mathematical models that are run on a digital computer.

Computer simulation is a powerful tool that can play a major role in the analysis and assessment of transportation facilities, especially when other analytical approaches are inappropriate. There are several advantages behind using simulation to study complex transportation systems. First, simulation allows for conducting what-if scenarios in a controlled environment without having to go out in the field and try different strategies that might not work. Second, simulation is much faster and more efficient than field experiments. It also results in a wealth of data on several system performance measures; data that would otherwise be impossible or impractical to collect in the field.

2.6.1 Traffic Simulation Models

There are several classification systems that could be used to categorize traffic simulation models. The simplest classification scheme would classify models based on the type of transportation facility they are designed to simulate. According to this classification, models would be divided into those for intersections for freeways and for urban networks. Another classification system would classify models into *stochastic* models and *deterministic* models. Stochastic models attempt to model the uncertainty associated with traffic phenomena through the use of statistical probability distribution functions. Deterministic models, on the other hand, assume that traffic flow is deterministic; while the assumption of a deterministic model is obviously not true, it simplifies the development and use of the model.

The most common classification scheme, however, classifies traffic simulation models based on their level of aggregation into microscopic, macroscopic, and mesoscopic simulation models. Microscopic models model individual vehicles and their interactions and are capable of simulating traffic to a high level of detail, but they do require intensive data input and a long execution time. Microscopic models are based on detailed car-following and lane-changing theories. Macroscopic models, on the other hand, model traffic as an aggregate fluid flow. They are based on the use of continuum models, similar to the Greenshields model, representing the relationship between speed, density, and flow. While macroscopic models are less detailed than microscopic models, they can work with larger networks and require less time to run.

Finally, mesoscopic models represent the middle ground between microscopic and macroscopic models. For example, while macroscopic models typically do not simulate lane-changing, merging, and diverging behaviors, mesoscopic models often do.

2.6.2 Examples of Traffic Simulation Models

During the last 20 or 30 years, a large number of sophisticated traffic simulation models were developed that are capable of simulating different traffic operations. A complete review of these models is beyond the scope of this chapter. Nevertheless, this section will review two of the most widely used traffic simulation models, particularly in the context of ITS operations; the two models are Corsim and Integration.

2.6.2.1 Corsim

Corsim is a traffic simulation model capable of simultaneously modeling surface streets as well as freeway systems [9]. The model represents a combination of two earlier models, Netsim and Fresim. Netsim is a microscopic model for modeling surface streets. In Netsim, individual vehicles are moved through the network based on car-following rules. The model is capable of modeling lane-changing behavior, turning movements, and response to traffic control devices. Fresim, on the other hand, is concerned with modeling freeway systems. It is also a microscopic model capable of simulating complex freeway geometrics and modeling various operational scenarios. Corsim provides the user with a wealth of output data detailing several systemwide, as well as link-specific, performance measures. The Corsim model has undergone extensive validation and evaluations since its development. It is currently one of the most widely used traffic simulation models, particularly in the United States.

2.6.2.2 Integration

Integration, on the other hand, is a model developed in the 1980s by Van Aerde and Associates. It is among the first models developed to model ITS operations and, in particular, real-time route guidance. It is a mesoscopic model in the sense that individual vehicles are traced through the network, but the details of lane-changing and car-following behavior are not modeled. Instead, the model moves the vehicles based on flow-density relationships similar to Greenshields model.

2.6.3 Basic Guidelines for Applying Simulation Models

The first step in applying simulation to a given transportation scenario is to select the appropriate simulation model. Simulation models vary in their attributes, strengths, and weaknesses, and therefore, it is quite important to select the right model. Once a model is selected, the next step is to calibrate the model to real-world observations. Calibration involves adjusting the model's parameters to get its results closer to reality. Finally, it is important to keep in mind that most simulation models are stochastic in nature, which means that the results will vary from one run to another based on the sequence of random numbers generated for each run. To address this variability, a simulation model should be run for an adequate number of times, and the means and variances of the different model parameters should be determined.

2.7 Conclusions

In this chapter, we introduced some of the basic concepts of traffic flow and control, which are quite useful for a full appreciation of several ITS applications. Specifically, the chapter introduced the three primary elements needed for describing a traffic stream (flow, density, and speed). Traffic models were discussed, as well as the phenomenon of shock waves in traffic streams and its applications. The basic principles of traffic signals and ramp meters were addressed, followed by a brief description of computer simulation and its application in traffic modeling. Equipped with this basic understanding, Chapter 3 describe what ITS can do for the different users of the transportation system.

Review Questions

1. What is the difference between free-flow speed and optimum speed?

2. Data obtained from an arterial photograph shows 10 vehicles occupying a 900-foot-long stretch of road. An observer standing by the side of the road also noted that vehicles pass by at approximately 3.0-second intervals. Determine: (a) the density on the road; (b) the flow on the road; and (c) the average speed.

3. How can the speed be represented on the fundamental traffic flow diagram [i.e., the diagram relating the flow (q) to the density (k)]?

4. Distinguish between the uses of the following three relationships of traffic flow: (a) the relationship between speed and density; (b) the relationship between speed and flow; and (c) the relationship between flow and density.

5. Assuming that the relationship between the speed and density is given by the following model:

$$u = u_f e^{-k/k_j}$$

Derive expressions for the speed and density at maximum flow (u_0 and k_0).

6. Explain what is meant by a shock wave in the context of traffic flow.

7. Given that the relationship between speed (u) in miles per hour and density (k) in vehicles per mile, obtained from actual data is

$$u = 54.5 - 0.24k$$

Determine: (a) the jam density; (b) the free flow speed; (c) the maximum capacity; and (d) the shock wave between conditions $u_a = 50$ mph and $u_b = 30$ mph.

8. A line of traffic moving at a speed of 40 mph and a density of 45 vph is stopped for 25 seconds at a red light. Assuming a jam density of 230 vph and a Greenshields model, determine (a) the length of the line of cars stopped during the 25 seconds of red, and (b) the number of cars stopped during the 25 seconds of red.

9. Distinguish between the following different modes of signals operation: (a) pretimed; (b) semiactuated; (c) fully actuated; and (d) computer-based control.

10. In the context of actuated signals, what is meant by the following terms: (a) minimum green; (b) maximum green; and (c) passage time interval?

11. What are the different types of ramp control?

12. Discuss the different types of traffic detectors that you would commonly find in a ramp metering system.
13. Distinguish between demand-capacity control and occupancy control strategies for traffic responsive ramp meters.
14. Discuss some of the basic guidelines for the use and application of traffic simulation models.

References

[1] May, A. D., *Traffic Flow Fundamentals*, Upper Saddle River, NJ: Prentice Hall, 1990.

[2] Garber, N. J., and L. A. Hoel, *Traffic and Highway Engineering*, 3rd ed., Pacific Grove, CA: Brooks/Cole, 2002.

[3] Papacostas, C. S., and P. D. Prevedouros, *Transportation Engineering and Planning*, Upper Saddle River, NJ: Prentice Hall, 2001.

[4] Roess, R. P., W. R. McShane, and E. S. Prassas, *Traffic Engineering*, 2nd ed., Upper Saddle River, NJ: Prentice Hall, 1998.

[5] Kell, J. H., and I. J. Fullerton, *Manual of Traffic Signal Design*, 2nd ed., Washington, D.C., Institute of Transportation Engineers: 1998.

[6] Kell, J. H., et al., *Traffic Detector Handbook*, 2nd ed., U.S. Department of Transportation/Federal Highway Administration, Publication No. FHWA-IP-90-002, July 1990.

[7] Transportation Research Center, *TRANSYT-7F Users Guide*, University of Florida, Gainesville, FL, 1998.

[8] U.S. Department of Transportation, Federal Highway Administration, *Freeway Management Handbook*, Report No. FHWA-SA-97-064, 1997.

[9] ITT Systems & Sciences Corporation, *CORSIM User's Manual*, Colorado Springs, CO, 1998.

3

ITS User Services

This chapter describes the different ITS user services. Within the context of ITS, the term *user services* describes what ITS should do for the users of the transportation system, including travelers, operators, planning organizations, state departments of transportation, and commercial vehicle operators, among others. To date, 32 ITS user services have been defined and grouped into 8 major user services bundles. This chapter discusses what is meant by the ITS user services and describes each of the 32 user services.

3.1 ITS User Services

In the early 1990s, there was recognition of the need to develop a framework to guide the deployment of ITS in the United States. To provide this framework, the National ITS Program Plan was developed [1]. The plan's primary focus was to define a list of interrelated services that ITS will be expected to provide. This list, which came to be known as the ITS user services, was compiled through a process that reflected the ideas and needs of a broad range of ITS stakeholders. At the time of this writing, 32 ITS user services have been defined. It is important to note, however, that this list will keep changing over time, as our understanding of ITS and what it can do for us improves. The original 29 user services and corresponding user service requirements are documented in the National ITS Program Plan. Subsequent addendums to the National ITS Program Plan include Highway-Rail Intersection, Archived Data Mangement, and Maintenance and Construction Operations.

The 32 ITS user services share a number of basic characteristics. First, the different user services can be regarded as building blocks for a region's intelligent transportation system. They can be combined for deployment in a number of different fashions, depending on a region's priorities and needs. Second, user services are made of multiple technological elements that may be common to other user services. A prime example of a technological element that many user services have in common is an advanced communications system for transferring data among the different transportation system components. Third, since user services share a number of multiple technological elements, the costs and benefits of a user service will depend on the deployment scenario. For example, once the basic technological elements (e.g., communications and surveillance) are in place, deploying a new user service may require only a small incremental cost. Finally, it should be noted that user services are not specific to a particular location. Rather, they can be adapted to meet the local needs of urban, suburban, or rural areas.

3.2 ITS User Services Bundles

The National ITS Architecture, which will be discussed in Chapter 5, groups the 32 user services into 8 bundles that can be regarded as focus areas for ITS applications. The grouping of the user services into these bundles was based on a number of criteria. In some cases, the grouping was based on the institutional perspective of the organization that will deploy the service; for example, a bundle was defined to group the services related to public transportation operations. In other cases, user services that share some common technical functions were grouped together, such as the advanced vehicle safety systems bundle. Table 3.1 shows the eight bundles, along with the user services making up each bundle. Section 3.3 will briefly describe each of these services [1].

3.3 Travel and Traffic Management

The main goal of the 10 user services grouped under this bundle is to use real-time information on the status of the transportation system to improve its efficiency and productivity, as well as to mitigate the adverse environmental impacts of the system. The main reason behind grouping these 10 user services together is to share information. As will be discussed, many of these user services share information with one another. This means that the real benefit behind ITS deployment would become much more obvious if these services are deployed in concert, resulting in a highly integrated system that would improve the efficiency, productivity, and environmental soundness of the existing surface transportation system. The following sections describe each of these 10 user services in more detail.

Table 3.1
User Services

Bundle	User Services
1. Travel and Traffic Management	1.1 Pretrip Travel Information
	1.2 En Route Driver Information
	1.3 Route Guidance
	1.4 Ride Matching and Reservation
	1.5 Traveler Services Information
	1.6 Traffic Control
	1.7 Incident Management
	1.8 Travel Demand Management
	1.9 Emissions Testing and Mitigation
	1.10 Highway-Rail Intersection
2. Public Transportation Operations	2.1 Public Transportation Management
	2.2 En Route Transit Information
	2.3 Personalized Public Transit
	2.4 Public Travel Security
3. Electronic Payment	3.1 Electronic Payment Services
4. Commercial Vehicle Operations	4.1 Commercial Vehicle Electronic Clearance
	4.2 Automated Roadside Safety Inspection
	4.3 Onboard Safety Monitoring
	4.4 Commercial Vehicle Administrative Processes
	4.5 Hazardous Material Incident Response
	4.6 Commercial Fleet Management
5. Emergency Management	5.1 Emergency Notification and Personal Security
	5.2 Emergency Vehicle Management
6. Advanced Vehicle Safety Systems	6.1 Longitudinal Collision Avoidance
	6.2 Lateral Collision Avoidance
	6.3 Intersection Collision Avoidance
	6.4 Vision Enhancement for Crash Avoidance
	6.5 Safety Readiness
	6.6 Precrash Restraint Deployment
	6.7 Automated Vehicle Operation
7. Information Management	7.1 Archived Data Function
8. Maintenance and Construction Management	8.1 Maintenance and Construction Operations

Source: [1].

3.3.1 Pretrip Travel Information

The goal of this user service is to provide travelers with information about the status of the transportation system *before* they begin their trip (hence the name, pretrip). The information provided is typically multimodal in nature and could include the following items:

1. Real-time flow conditions (e.g., average traffic speeds or travel time);
2. Road incidents and suggested alternate routes;
3. Scheduled road construction and any special events;
4. Transit routes, schedules, fares, and transfers;
5. Park-and-ride facilities locations and availability.

Travelers will be able to access this information at home or work (e.g., via a computer or a telephone system) and at major trip generators (e.g., using a touch-screen kiosk at a shopping mall). As these pretrip information systems mature, they would be provided with the capabilities needed to plan trip itineraries and to provide travelers with mode choices based on real-time travel conditions and their preferences. The intent of this user service is to provide travelers with enough information to help them make more informed decisions regarding their time of departure, the mode to use, or the route to take to their destination.

3.3.2 En Route Driver Information

As opposed to pretrip travel information, this service is aimed at providing drivers with travel-related information *after* they start their trip (en route). The service can be regarded as consisting of two subservices: the driver advisory subservice and the in-vehicle signing subservice. The goal of the driver advisory subservice is to provide travelers with information similar to that provided by the pretrip information service but, in this case, information is provided en route using devices such as *variable message signs* (VMSs), the car radio, or *portable* communications devices. By providing drivers with real-time information, this service hopes to better utilize the existing capacity of the highway network. For example, drivers may be able to switch to an alternate route once informed of an incident or extreme congestion ahead.

The goal of the in-vehicle signing service, on the other hand, is to provide drivers with *in-vehicle* displays of roadway signing and warning messages such as stop signs, sharp curves, reduced speed advisories, and so on. This information would be presented to drivers in various media to accommodate any special needs that they might have. For example, the information could be presented as a voice output to overcome a visual impairment or as a head-up display to overcome a hearing impairment. The goal of this service is to improve the safety of operating a vehicle. Applications of in-vehicle signing could include warning motorists of unsafe conditions such as sharp curves, wet pavements, and icy conditions; alerting motorists if they exceed the safe speed limit; and alerting drivers of unsafe weather conditions.

3.3.3 Route Guidance

This user service is intended to provide drivers with detailed turn-by-turn instructions on how to get to their destinations. These directions could be based on static information (i.e., historical travel times for the different road segments) or on real-time information, which would take into account current traffic speeds and current incidents' locations. As can be seen, this service is closely related to the en route driver information service. However, while the en route service just provides a driver with travel information and leaves the final selection of the route to the driver, the route guidance service processes this information into detailed navigational directions. Users of the route guidance service are likely to include drivers of private vehicles, commercial vehicle operators, and public transit drivers (particularly demand responsive transit). The benefits of this service include reduced travel delay through the reduction of wasted travel time as a result of navigational errors, as well as decreased driving stress levels, especially when driving in an unfamiliar area.

3.3.4 Ride Matching and Reservation

This user service attempts to encourage carpooling by providing real-time matching of the preferences and demands of users with providers and by serving as a clearinghouse for financial transactions. Under this service, travelers would call a service center and provide the center with information regarding the desired trip origin, destination, and time. Travelers would then receive a number of ridesharing options from which they could choose.

3.3.5 Traveler Services Information

This service is intended to provide travelers with Yellow-Pages-type information, including the location of services such as food, lodging, gas stations, hospitals, and police stations, as well as the location of points of attraction for tourism purposes. The information will be made available to travelers both at home or work (as an add-on feature to a pretrip information system), as well as en route (in the vehicle or at public facilities such as rest areas).

3.3.6 Traffic Control

This user service is aimed at managing and controlling the movement of traffic on streets and highways in an attempt to optimize the use of such facilities. The service encompasses adaptive signal control systems along arterials that adjust signal timings to current traffic conditions, as well as freeway control systems that use ramp metering and lane control techniques to smooth traffic flow. It also includes the integration of freeways and network signal systems for the purpose of areawide optimization of traffic flow.

The traffic control user service is responsible for collecting real-time traffic data from the field, processing the data into useful information, and using the information to assign the right of way to transportation users in the most efficient fashion. The service can be regarded as one of the most basic building blocks of an ITS, since its detection, control, communications, and support systems are fundamental to the operation of several other ITS services. The physical manifestation of this user service can currently be observed at almost every major city in the United States in the form of several traffic operations centers that exist throughout the country.

3.3.7 Incident Management

This user service aims to improve the existing capabilities for detecting and responding to incidents. This includes both unpredicted incidents (e.g., a car accident), as well as planned incidents (e.g., scheduled road construction). The main idea here is to use advanced sensors (such as closed-circuit TV cameras), data processing, and communications technologies to quickly detect and verify an incident. Sophisticated decision support systems will then help operators reach the best set of actions needed to appropriately respond to the scenario involved. Studies clearly show that reducing the time needed to detect, verify, and clear incidents has the potential to result in tremendous savings to highway users and state agencies alike. It should be noted that this service is closely related to the traffic control service. After the incident management service detects an incident, the traffic control service would need to take the appropriate measures to mitigate the impacts of such an incident.

3.3.8 Travel Demand Management

The main purpose of the travel demand management user service is to develop and implement strategies aimed at reducing the number of *single-occupancy vehicles* (SOVs), while encouraging the use of *high-occupancy vehicles* (HOVs). The service also strives to provide travelers who would opt for a more efficient travel mode with several mobility options. The service includes the following strategies:

- *HOV facility management and control.* The goal here is to operate HOV facilities in a fashion that is responsive to current conditions. For example, occupancy requirements could be adjusted throughout the day based on current traffic and congestion levels.
- *Congestion pricing.* This involves adjusting tolls to encourage mode changes and reduce travel demand. For example, tolls could be increased during peak hours in urban areas or around environmentally sensitive tourist attractions in rural areas.
- *Parking management and control.* The idea here is to dynamically manage the allocation and price of parking spaces to effect a mode change to HOVs. For example, parking fees for SOVs could be increased during peak periods, while offering discounts to HOVs.
- *Mode change support.* This strategy is intended to support the ride matching and reservation user service previously described (Section 3.3.4). The goal here is for travelers to be able to call a *traffic management center* (TMC) with information about their destination and the desired time of departure. The TMC would then find a carpool willing to take additional riders.
- *Telecommuting and alternative work schedules.* The goal here is to take advantage of recent advances in telecommunications technologies and alternative work schedules to avoid driving to and from work during peak periods.

3.3.9 Emissions Testing and Mitigation

This user service is based on the use of environmental sensors to collect information about exhaust emissions from specific vehicles, at a certain location, or over a wide area. The information collected can then be used in a number of ways. For example, the information could be used to divert traffic away from sensitive air quality areas or to control access to these areas. The information could also provide valuable insight for developing air quality improvement strategies. Finally, it could be used to alert vehicles' operators if their vehicles are not in compliance with adopted emissions standards.

3.3.10 Highway-Rail Intersection

The purpose of this user service is to provide for improved warning and control devices at *highway-rail intersections* (HRIs). The control and warning devices at HRIs will be interconnected with adjacent signalized intersections so that local control can be adapted to HRI activities. The service will also provide for monitoring the "health" of HRI equipment and reporting any detected malfunctioning.

3.4 Public Transportation Operations

Four user services are included under this second service bundle. As the name implies, these services are concerned with improving the service of public transportation systems in an attempt to encourage their use. The following sections describe each of these four user services.

3.4.1 Public Transportation Management

This service uses advanced communications and information systems to collect data that can be used to improve the operations of vehicles and facilities; the planning and scheduling of service; and the management of personnel. The following discusses each of these three application areas:

1. *Improving the operations of vehicles and facilities:* In this application, real-time data from the vehicles (e.g., data about the vehicle's location) is communicated via a digital data link to the transit center (see Figure 3.1). At the transit center, the real-time location of the vehicle is compared against schedule information. Schedule deviations are then identified, and corrective actions for returning the vehicles to schedule are determined and communicated to drivers. The real-time vehicles' location information can also be used to facilitate transfers by ensuring that planned vehicle meetings do indeed take place.

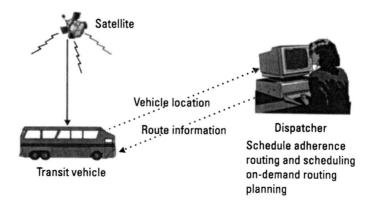

Figure 3.1 Improving the operations of transit vehicles.

2. *Improving planning and scheduling:* This application involves the off-line storage and analysis of the real-time data being collected, including data on passenger loading, bus running times, and mileage accumulated by vehicles. These data could be used to revise schedules, to plan routes, and to satisfy federal reporting requirements. Moreover, the data could be used to feed customer information systems.

3. *Personnel management:* This application is concerned with assigning drivers to routes based on their seniority and route preferences. It also involves assigning service technicians by skill level and automating the generation of periodic maintenance schedules by monitoring the daily miles traveled by each vehicle.

3.4.2 En Route Transit Information

This service is intended to provide transit riders with information *after* their trips have started, including information on expected arrival times of transit vehicles, transfers, connections, and ride-share opportunities. The goal is to assist travelers in making effective transfer decisions, as well as to increase the level of comfort and convenience of the trip. The information would be provided through interactive kiosks at key locations, interactive displays on buses, and visual displays at bus stops.

3.4.3 Personalized Public Transit

Personalized public transit (PPT) centers around the idea of using flexibly routed vehicles to offer more convenient service to travelers (i.e., door-to-door service). There are basically two types of PPT: flexibly routed operations and random route operations. In flexibly routed operations, fixed-route buses are allowed to deviate from their main route to pick up or drop off passengers. Random route operations, on the other hand, operate on a totally random route that is determined based on the service requests received.

At the present time, most systems (with the exception of taxis) require passengers to make their reservations at least 24 hours in advance, with "instantaneous" requests honored only if there is enough slack in the schedule. The goal of this user service is to allow reservations, vehicle assignments, and scheduling to be developed in real time. Under this service, travelers would provide a dispatch center with information on their trip origin and destination. A computer at the center would then assign the closest vehicle to service the request and would automatically inform the driver of that vehicle of the passenger's origin and destination. The system would also notify the traveler when to expect the vehicle's arrival to reduce passenger anxiety and uncertainty.

3.4.4 Public Travel Security

This user service uses advanced technologies to improve the security of public transportation through the detection, identification, and notification of security incidents. Under this service, transit stations, parking lots, and bus stops will be provided with closed-circuit television systems for surveillance, as well as with push-button alarms directly connected to central dispatch or to the police. The vehicles will also be equipped with onboard silent alarms and connected microphones for monitoring. Finally, for real-time ride-matching systems, participants may be identified through computerized identification cards.

3.5 Electronic Payment

This service bundle consists of only one user service, namely the *electronic payment services* (EPS). The main goal here is to provide travelers with a common electronic payment medium for different transportation modes and functions. EPS consists of four main components:

- *Electronic toll collection (ETC)*. These systems allow drivers to pay tolls without stopping or slowing down from cruising speed, thereby minimizing delays and travel times, as well as improving air quality in the vicinity of toll plazas. ETC systems could take the form of a main line barrier plaza, an open highway collection system where tolls are collected at main line speeds, or a closed system where tolls are based on entry and exit points. These systems will also be able to determine tolls for different classes of vehicles, confirm toll charges through the use of roadside and in-vehicle display devices, and provide for the automatic enforcement of violations.

- *Electronic fare collection*. These systems will relieve transit users from the need to provide exact cash for payment through the use of fare cards (i.e., smart cards). They will also allow for implementing more complex fare structures and facilitate the participation of employers in transit benefit programs. Finally, electronic fare collection systems will help transit operators in the areas of operations and route planning by providing more accurate data about passenger boardings.

- *Electronic parking systems*. These systems are intended to enable drivers to pay for parking without cash through the use of electronic cards or tags.

- *Integrated EPSs*. The ultimate goal of EPSs is to integrate the previously mentioned components to enable one payment medium to be accepted for all transportation services in a region. With an integrated payment system, a traveler driving on a toll road, headed to a park-and-ride lot to use transit would be able to use the same device to pay for toll, parking, and the transit fare.

3.6 Commercial Vehicles Operations

As the name implies, the user services falling under this bundle aim at improving the efficiency and safety of *commercial vehicles operations* (CVOs). As will be discussed, these user services have two main focus areas: improving private-sector fleet management and streamlining government/regulatory functions. The following sections briefly describe each of the six CVO user services included under this bundle.

3.6.1 Commercial Vehicle Electronic Clearance

This user service will allow complying commercial vehicles to continue past checkpoints at main line speeds. As a vehicle approaches a checkpoint, communications between the vehicle and the inspection station will take place, allowing authorities to check the vehicle's credentials, weight, safety status, and cargo. Enforcement personnel can then select potentially unsafe vehicles for inspection, while permitting safe vehicles to bypass the checkpoint.

3.6.2 Automated Roadside Safety Inspection

This user service will use automated inspection capabilities to allow for checking safety requirements more quickly and more accurately during a safety inspection at an inspection site. Innovative devices will be used to check brake performance, steering, and vehicle suspension systems performance, as well as to assess the driver's current performance and alertness. The use of these advanced capabilities will reduce the amount of time spent for inspection, while providing a more accurate picture of the safety status of the vehicle.

3.6.3 Onboard Safety Monitoring

This user service provides for the capability to monitor the safety status of a vehicle, cargo, and driver at main line speed and is therefore different from the automated roadside safety inspection, which assesses safety status after a vehicle has pulled off the highway. The service will be integrated with the commercial vehicle clearance service to allow for reading out this safety status at main line speeds and with the automated roadside safety inspection service to allow enforcement personnel to read out this safety status at the roadside. The monitoring capabilities will include the following:

- Sensing the condition of critical vehicle components such as brakes, tires, and lights;
- Sensing shifts in cargo while the vehicle is in motion;
- Monitoring time-on-task for the drivers;
- Monitoring drivers' alertness levels.

Safety warning would be provided to drivers and be made available for carriers and roadside enforcement personnel.

3.6.4 Commercial Vehicle Administrative Processes

This user service will give carriers the ability to purchase credentials automatically and with the functionality needed for automated mileage and fuel reporting. The idea here is to streamline government and regulatory functions in order to save carriers time and money. Both components are described in further detail next.

The electronic purchase of credentials component will allow carriers to file applications electronically for credentials such as registration, fuel use taxes, trip permits, oversize or overweight permits, or hazardous materials permits. The credentials will be approved in a much shorter time than is possible with the current paper system (i.e., in hours versus weeks). This component will also allow for electronic data interchange and electronic funds transfer capabilities for the movement of data and funds between agencies and carriers. Systems will be developed to allow for one-stop shopping, whereby carriers would be able to obtain permits for multiple states through a single source.

Regarding the automated mileage and fuel reporting component, according to federal and state regulations, an interstate carrier is required to collect, report, and maintain accurate mileage and vehicle information for each trip by state. Registration fees and fuel taxes that the carrier has to pay are then based on the proportion of miles traveled in each state during the previous year, with the amount of fuel tax paid by a carrier with each fuel purchase in a particular state deducted from the tax as calculated by the mileage.

This regulation process represents a huge administrative burden on carriers (in collecting and reporting mileage and fuel information), as well as on states (in processing this information). The automated mileage and fuel reporting and auditing component will allow carriers to automatically record the miles traveled and fuel purchased in each state. This data could then be downloaded, compiled, and submitted automatically to the states involved. Capabilities will be developed for the states to audit the automatic recording and reporting.

3.6.5 Hazardous Materials Incident Response

Responding to hazardous materials incidents requires a full knowledge of the type of materials involves. The primary goal of this user service is to provide law enforcement and hazardous materials response personnel with timely, accurate information on cargo contents at the scene of an accident, so that they can know exactly how to handle the materials involved in an appropriate fashion. This information could reside in the infrastructure (e.g., existing carrier databases, state information systems) or in vehicle-based systems (e.g., on vehicle-mounted transponders). Emergency responders will be provided with access to this information either through remote access to the pertinent databases or through the use of readers that could communicate with transponders onboard the vehicles.

3.6.6 Freight Mobility

This user service will provide real-time communications for vehicle location, dispatching, and tracking between drivers, dispatchers, and intermodal transportation providers.

3.7 Emergency Management

Two user services are included under this bundle, namely emergency notification and personal security and emergency vehicle management. Each user service is discussed in the following sections.

3.7.1 Emergency Notification and Personal Security

There are two main components to this user service: driver and personal security and automated collision notification. The driver and personal security subservice will provide users with the ability to *manually* initiate a distress call for emergency and nonemergency incidents such as mechanical breakdowns. The system providing this functionality will have the capability to automatically determine the vehicle's location and to transmit the location as a part of the notification message.

The automated collision notification subservice, on the other hand, is designed to *automatically* notify *emergency management services* (EMS) personnel of serious automobile accidents. The systems providing this subservice will have the ability to automatically detect the occurrence of an incident, automatically identify the location of the incident, and automatically relay a notification message to an EMS dispatcher that includes the location of the incident and its severity.

3.7.2 Emergency Vehicle Management

The main goal of this user service is to reduce the time from the receipt of an emergency notification to the arrival of the vehicle at the scene of the incident, thereby reducing the severity of crashes injuries. Reducing the response time will be accomplished through the utilization of the following three subservices:

1. *Emergency vehicle fleet management.* This subservice will provide the functionality needed to identify the locations of emergency vehicles in real-time to allow for dispatching the vehicle that can most swiftly reach the incident scene.
2. *Route guidance.* This subservice is concerned with helping the emergency vehicle driver determine the fastest route to the incident scene or to a suitable hospital.
3. *Signal priority.* This service will provide for preempting traffic signals for emergency vehicles, as well as the capability of warning drivers of an approaching emergency vehicle.

3.8 Advanced Vehicle Control and Safety Systems

The user services included under this bundle are mainly intended to improve the safety of the transportation system by supplementing drivers' abilities to maintain vigilance and control of the vehicle and by enhancing the crash avoidance capabilities of motor vehicles. The significance of attempting to supplement drivers' abilities should be quite apparent, considering the fact that statistics show that driver error is a major factor in more than 90% of car accidents in the United States. The following seven user services are included under this bundle.

3.8.1 Longitudinal Collision Avoidance

This service will help drivers avoid, or decrease the severity of, longitudinal crashes (i.e., crashes where the vehicles' paths prior to the collision are parallel, including rear-end, backing, and head-on types). To accomplish this, the following four systems will be implemented:

1. Rear-end crash warning and control;
2. *Adaptive cruise control* (ACC);
3. Head-on crash warning and control;
4. Backing crash warning and control.

These four systems will assist drivers by sensing potential hazards to the front or rear of the vehicle, warning of impending collisions, and providing temporary automatic control of the vehicle to avoid collision situations:

1. *Rear-end collision avoidance.* Rear-end collisions generally occur as a result of a driver's failure to maintain a sufficient separation distance from the vehicle in front or failure to perceive a slow or stopped vehicle ahead. Rear-end collision avoidance systems can monitor this separation distance and warn drivers of any sensed dangerous situation. If the driver does not react in an appropriate fashion to the sensed danger, automatic vehicle control actions might be initiated.

2. *ACC.* These systems will track the vehicle in front and automatically maintain a desired minimum distance from that vehicle. When this minimum distance is maintained, a vehicle would travel at the set speed. However, if the separation distance falls below this minimum value, an ACC system would warn the driver or perhaps initiate control actions (such as downshifting or braking) to slow the vehicle and regain the minimum headway distance.

3. *Head-on collision warning and control.* These systems detect impending collisions with vehicles moving in the opposite direction and in the same lane as the equipped vehicle. In order to be able to implement appropriate actions, these systems will need to have a complete picture of the road configuration, adjacent vehicles, and roadside hazards. Given this, it is most likely that these systems will require vehicle-to-vehicle communications.

4. *Backing collision warning and control.* These systems will detect slow moving or stationary objects or pedestrians that are in the path of a backing vehicle and warn the driver accordingly. Detecting these objects will require the use of relatively short-range sensors onboard the vehicles.

3.8.2 Lateral Collision Avoidance

These systems aim at helping drivers avoid accidents that result when a vehicle leaves its own lane of travel by warning drivers and/or assuming temporary control of the vehicle. The following subservices are included under this user service:

- Lane change/blind spot situation display, collision warning, and control;
- Lane/road departure warning and control.

These two subservices are described as follows:

1. *Lane change/blind spot situation display, collision warning, and control.* These systems will provide drivers with information about the presence of vehicles in their blind spots, warn them of potentially dangerous lane change maneuvers, and might even assume temporary control of the vehicle's steering and braking to avoid collisions. The systems providing this functionality could be implemented in three stages. At the first stage, the systems will provide continuous information about the presence of any vehicle in the driver's blind spots. At the second level, the systems will warn drivers of potential collisions when initiating a lane-changing maneuver. In this case, the systems will be activated by a cue that the driver is intending a lane change (e.g., using the turn signal). Finally, the third stage will involve the system temporarily assuming control of the vehicle systems, such as steering and braking.

2. *Lane/road departure warning and control.* The purpose of these systems is to assist in keeping a vehicle in its proper lane of drivers through driver warnings or assuming temporary control of the vehicle. These systems will detect the road or lane edge using either vehicle-based technologies (such as video image processing and optical or infrared scanning) or using vehicle-infrastructure technologies (such as vehicle-based radar units along with reflectors mounted along the edge of the road or lane).

3.8.3 Intersection Collision Avoidance

The primary goal of this user service is to avoid, or to decrease the severity of, collisions at intersections. Two of the most common types of collisions occurring at intersections are *straight crossing path* (SCP) collisions, which occur when two vehicles attempt to pass *simultaneously* through an intersection at right angles; and *crossing left turn* (CLT) collisions, which occur when a vehicle is attempting to turn left against opposing traffic. This user service aims at helping drivers avoid these types of collisions.

The system will track the position and state of vehicles within a defined area surrounding an intersection through the use of vehicle-to-vehicle communications or vehicle-to-infrastructure communication. For example, if a driver is attempting to cross another main road, the driver could be alerted if there is a fast vehicle approaching the intersection. In addition, once the driver begins crossing the intersection, other vehicles could be warned of the crossing vehicle.

3.8.4 Vision Enhancement for Collision Avoidance

This user service is intended at eliminating or reducing the severity of accidents where impaired or reduced visibility (such as night or foggy conditions) is the main causal factor. The service will be implemented through in-vehicle sensors capable of capturing an image of the driving environment and providing a graphical display of the image to the driver (e.g., through a head-up display). This improved visibility will help drivers avoid potential collisions with other obstacles, as well as help them better comply with traffic control signals.

3.8.5 Safety Readiness

The goal of this user service is to reduce the number of crashes caused by impaired drivers, failure of vehicles' components, and degraded infrastructure conditions. Accident statistics clearly show that a significant percentage of crashes occur as a result of drivers intoxication or drowsiness, failure of a critical component of a vehicle (e.g., brakes or tires), and icy or snowy road surfaces. This user service will help address these causes through the following three subservices:

1. *Impaired driver warning and control override.* These systems will monitor the performance of the driver and either warn or assume temporary control of the vehicle in case a driver's performance is impaired.
2. *Vehicle condition warning.* These systems will continuously monitor the performance of the critical components of a vehicle and warn drivers of their imminent failure.
3. *In-vehicle infrastructure condition monitoring.* These systems will monitor the roadway and provide warning to the driver of unsafe conditions such as loss of tire traction because of wet or icy road surfaces.

3.8.6 Precrash Restraint Deployment

This user service is designed to anticipate an imminent collision and activate the appropriate passenger safety systems prior to the actual impact. Current passenger safety systems (e.g., air bags) are activated *after* the onset of the crash. The precrash restraint deployment service plans to improve on existing systems by attempting to *anticipate* an imminent crash through the use of sensors capable of detecting the rapid closing of distance between the vehicle and an obstacle. In addition, this user service is designed to help reduce the danger of side impact collisions by setting a restraint to absorb or dissipate the force of the impact.

3.8.7 Automated Highway System

This user service has the ambitious goal of developing a fully *automated highway system* (AHS) where specially equipped vehicles will travel, under fully automated control, along dedicated highway lanes. The AHS concept has the potential to significantly improve the safety, as well as the efficiency, of highway travel through reducing the number and severity of crashes, decreasing congestion, and reducing vehicle emissions and fuel consumption. The vision for an AHS is as follows. Before entering an AHS, a vehicle would have to go through a checkpoint where a system would check the adequacy of the vehicle and the driver to use the AHS. If approved, the AHS would assume control of the vehicle and would move along the automated lanes until the destination exit is reached. At that point, the system would move the vehicle to an exit ramp and would return control to the driver after his or her ability to resume operation of the vehicle has been demonstrated.

3.9 Information Management

This service bundle is made up of a single user service, the Archived Data Function, commonly called the *Archived Data User Service* (ADUS). The purpose of this user service is to provide the functionality needed to store and archive the huge amounts of data being collected on a continuous basis by the different ITS technologies. The data collected can then be used for transportation planning, administration, and research purposes.

3.10 Maintenance and Construction Management

This service bundle is made up of a single user service, the *Maintenance and Construction Operations* (MCO) user service. The purpose of this user service is to provide the functionality needed for managing the fleets of maintenance vehicles, managing the roadway with regard to maintenance and construction, managing work zones and safe roadway operations, and the dissemination and coordination of roadway maintenance conditions and work plans. The scope of maintenance vehicles' fleet mangement includes systems that monitor and track maintenance vehicle location, support enhanced routing, scheduling and dispatching, and use of onboard diagnostic systems to assist in vehicle operations and maintenanace activities. The scope of roadway management includes systems that provide automated monitoring of traffic, road surface, and weather conditions, coordinated dispatching, hazardous road conditions remediation, and the ability to alert the public operating agencies of changing conditions. The scope of work zone management and safety includes systems that ensure safe roadway operations during construction and other work zone activities and includes communications with the traveler. The scope of the roadway maintenance conditions and work plan dissemination includes systems that disseminate and coordinate MCO work plans to affected personnel within and between public agencies and private sector firms.

3.11 Conclusions

In this chapter, we discussed the 32 user services in 8 different ITS user service bundles that describe what ITS can do for users of the transportation system. It should be noted, however, that as new technologies and systems are developed in the future, more ITS user services will have to be added, which will help broaden the scope and range of ITS applications. In Chapter 4, we will describe some real-world examples of ITS systems that implement the ideas we described in this chapter. Also, in Chapter 5, the relationship between the user services, thier correspoinding user service requirements, and the National ITS Architecture will be discussed.

Review Questions

1. Explain what is meant by an ITS user service.
2. What are some of the basic characteristics of ITS user services?
3. Select an ITS project in your state that you are familiar with and identify the user services upon which the project was based.
4. Explain the difference between the pretrip and the en route travel information user services.
5. Describe how real-time information about the location of transit vehicles could be used to improve the service.
6. In the context of PPTs, what is the difference between flexibly routed operations and random route operations?
7. What are some of the benefits of ETC systems?
8. Explain the rationale behind the automated mileage and fuel reporting user service falling under the CVO service bundle.
9. How does the emergency vehicle management user service attempt to reduce the time elapsed between the time of the receipt of an emergency notification and the time of the arrival of an emergency vehicle at the incident scene?

10. Describe the difference between longitudinal and lateral crash avoidance systems.
11. Describe what is meant by the automated highway concept and list some of its anticipated benefits.

Reference

[1] U.S. Department of Transportation, Federal Highway Administration, *The National Intelligent Transportation Systems Program Plan*, Washington D.C., 1995.

4

ITS Applications and Their Benefits

This chapter describes some real-world examples of ITS applications, along with their reported or anticipated benefits. Our intent is to provide a more tangible view of ITS, compared to the rather abstract nature of ITS user services described in Chapter 3, and to help the reader appreciate what ITS can and cannot do. The chapter also hopes to demonstrate that ITS is becoming a part of our national transportation system. Examples of ITS applications can now be found in almost all 50 states.

It is almost impossible to provide a comprehensive overview of all ITS applications, so we limited our discussion to applications that would normally be deployed within a metropolitan area. This, however, should by no means imply that ITS is only for metropolitan areas. In fact, there is a whole ITS program intended for rural areas called the *Advanced Rural Transportation System* (ARTS) program. In addition, many states have a comprehensive statewide ITS program, which includes both urban and rural elements.

Even with this focus, the examples presented in this chapter cannot be regarded as a complete overview of metropolitan ITS applications. They are primarily intended to give the reader a feel for some of the most successful ITS applications. The examples can be categorized into the following four application areas:

1. Freeway and incident management systems;
2. Advanced arterial traffic control systems;
3. Advanced public transportation systems;
4. Multimodal traveler information systems.

For each application area, the operational concept will be described first, followed by a list of some real-world examples and their associated benefits.

4.1 Freeway and Incident Management Systems

To combat freeway congestion, the concept of *freeway and incident management systems* (FIMS) was introduced. FIMS strive to support traffic flow by attempting to use existing capacity as efficiently as possible, because the traditional solution of more road building is no longer a viable option.

Given the critical importance of the freeway network to urban transportation systems, FIMS are typically among the first ITS components to be deployed in large metropolitan areas. According to the *Freeway Management Handbook* [1], freeway management can be defined as the control, guidance, and warning of traffic in order to improve the flow of people and goods on limited access facilities. To do this, FIMS combine field equipment (such as traffic detectors, variable message signs, and ramp meters), communications (such as interconnections with other centers and agencies), traffic control centers with their computer hardware and software, and the people who staff these centers. All these elements allow FIMS to control and manage traffic more efficiently in an effort to prevent congestion.

Congestion on a freeway occurs when demand exceeds capacity. There are generally two types of congestion, *recurrent* congestion and *nonrecurrent* congestion. Recurrent congestion typically occurs on a regular basis (e.g., during peak hours). It is the result of demand exceeding the normal capacity of a freeway during a certain period of time and can be quite predictable. Nonrecurrent congestion, on the other hand, is quite unpredictable and is the result of unusual occurrences such as traffic accidents, adverse weather conditions, and short-term construction work. These unusual events result in a reduction in the normal capacity of a freeway segment. This, in turn, results in congestion when the reduced capacity drops below a level that can accommodate travel demand. FIMS can deal with both types of congestion. They are more effective, however, in dealing with nonrecurrent congestion.

4.1.1 FIMS Objectives

While the overall goal of FIMS is to improve the flow of traffic over limited-access facilities, the following more specific objectives are commonly identified for FIMS:

1. To continuously monitor the status of traffic flow and to take appropriate traffic control actions aimed at smoothing traffic flow, when a need arises;

2. To reduce the frequency of recurrent congestion and to mitigate its adverse impacts;
3. To minimize the duration and severity of nonrecurrent congestion by quickly restoring capacity to its normal level;
4. To maximize freeway efficiency and to improve public safety;
5. To provide travelers with real-time travel information on the status of traffic to allow them to make more intelligent route and mode choices.

In order to achieve these objectives, FIMS combine and integrate a large number of elements and components. These elements can generally be divided into three groups: roadside elements, control center elements, and communications elements. Roadside elements may include different types of traffic detectors, closed-circuit TV cameras, and environmental sensors, as well as ramp meters for controlling traffic flow. They also include the means for disseminating information and communicating important messages to drivers, such as variable message signs and highway advisory radio. The control center, with its hardware and software components, serves as the central location for collecting, processing, and disseminating information on the state of the freeway system. It is also the place where operators, guided by *decision support systems* (DSSs), make management and control decisions aimed at optimizing the performance of the freeway facility. Finally, a communications system is required to link the control center to the field elements, as well as to link it to other regional control centers (e.g., emergency management centers).

4.1.2 FIMS Functions

Generally speaking, the functions performed by an FIMS can be broadly classified into the following functional areas, as shown in Figure 4.1:

- *Traffic surveillance and incident detection.* This function involves the continuous monitoring of traffic flow conditions and the driving environment, as well as the quick detection of any abnormal conditions such as accidents. Traffic surveillance provides the data needed to support the other functions, taking effective management and control actions.
- *Ramp control.* This involves controlling the flow rate entering the freeway system in order to maximize overall freeway operations efficiency, reduce traffic turbulence at ramp junctions, and improve traffic safety.
- *Incident management.* This includes all the actions needed to manage incidents, including incident verification, incident clearance, and site management, as well as traffic management during the incident (e.g.,

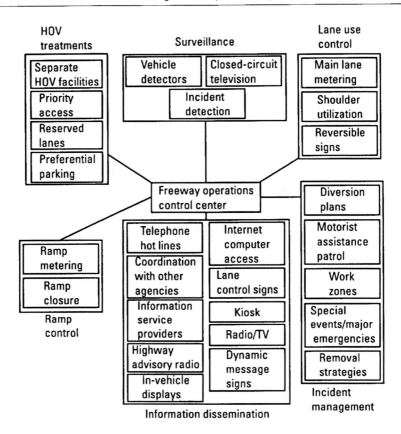

Figure 4.1 Functions of an FIMS. (*After:* [1].)

traffic diversion to alternate routes). The goal here is to quickly restore the freeway to its full capacity after an incident, as well as to mitigate the negative impacts of incidents as much as possible.

- *Information dissemination.* This function is responsible for disseminating real-time information on the status of traffic flow to the public, so that they can make more informed route and mode choice decisions. Information disseminated could include traffic incidents and possible alternate routes, scheduled highway construction, and expected weather conditions.
- *Lane use control.* This function is responsible for controlling and regulating the use of travel lanes and shoulders in an attempt to maximize the utilization of existing capacity. Examples of lane use control actions include opening shoulders to through-traffic during peak periods and reversible lane operations.

- *Providing preferential treatment to HOVs.* This function involves actions intended to encourage the use of HOVs by generating a time advantage for such vehicles. Examples of such actions include giving HOVs priority at ramp meters, using special surveillances to detect and remove incidents quickly from HOV lanes, and using technology to enforce HOV rules.

In the following sections, we discuss the most important functions in more detail and describe the different elements and technologies they require.

4.1.3 Traffic Surveillance and Incident Detection

The traffic surveillance function, the most basic and fundamental of all ITS functions, is responsible for gathering continuous, real-time information on the state of the transportation system and the driving environment. The information gathered forms the basis for the different traffic management and control actions that can be taken to optimize performance and also serves as the basis for the information to be disseminated to travelers.

The traffic surveillance system can be designed to serve a number of useful purposes that include the following:

1. To quickly detect and verify the occurrence of incidents on the freeway;
2. To monitor the clearance of incidents;
3. To provide real-time information on the status of traffic flow in support of the development of effective control strategies;
4. To monitor hazardous environmental conditions such as icy roads, flooding conditions, and high winds.

To support these functions, the traffic surveillance system will need to collect several types of data, the most important being the status of traffic operations. Historically, traffic operations are evaluated on the basis of the three fundamental measures of traffic flow discussed in Chapter 2 (flow, speed, and occupancy or traffic density). While these measures still constitute an essential component of the data collected by modern surveillance systems, the collection of additional types of data is now possible, thanks to recent advancements in surveillance technologies. These other types of data include the following:

- Real-time video images of transportation system operations;
- Queue length information;

- Travel time information between given origin-destination points;
- Real-time information on the location of emergency response vehicles;
- Real-time bus or transit vehicle location information;
- A host of environmental data, including pavement temperature, wind speed, and road surface condition information, along with information on emissions levels and air quality.

The information collected is typically communicated, on a real-time basis, to the freeway management system's operations center, where decisions regarding the control of the freeway system are made. We will now discuss in detail the different components that make up the surveillance system, as well as the different technologies that can be used to implement the system.

The surveillance system can be regarded as consisting of the following four basic components (see Figure 4.2):

1. Detection methods;
2. Computer hardware;
3. Computer software;
4. Communications.

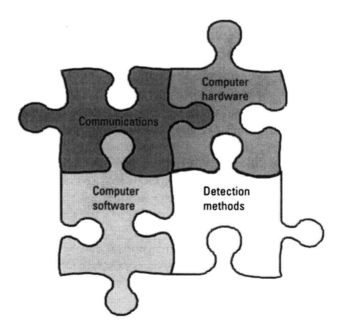

Figure 4.2 Surveillance system components.

Detection methods typically employ several different technologies, including inductive loops, nonintrusive detection devices, closed-circuit TV cameras, probe vehicles, and police and citizen reports. They also employ environmental sensors for monitoring weather-related conditions. The hardware elements include computers, monitors, controllers, and displays. The computer software elements play the major role of converting the data collected by the detection devices into meaningful information; they also allow for interfacing and communicating with the field devices. Finally, the communications system connects the different components of the control center together and links the center to the field devices. The following sections describe each of these components in more detail and discuss the different technologies available.

4.1.3.1 Detection Methods

As previously mentioned, a number of alternate technologies are available for collecting real-time traffic information. These include *inductive loop detectors* (ILDs) and nonintrusive detectors (e.g., microwave, infrared, ultrasonic, and acoustic sensors), *closed-circuit TV* (CCTV) cameras and *video image processing* (VIP), vehicle probes, police and citizens' reporting, and environmental sensors [1].

ILDs

ILDs are currently the most widely used devices for vehicle detection. Their main use is at intersections in conjunction with advanced signal traffic control systems, but they could also be used on freeways for incident detection purposes. ILDs typically take the form of one or more turns of insulted wire embedded in the pavement. The loop is connected via lead-in cable to the detector unit, which detects changes in the loop inductance when a vehicle passes over it. ILDs can be used to detect a vehicle's presence or passage. They can also be used to measure speed (by using two loops a short distance apart) and for classification. The main problem with using ILDs, however, is their reliability. Because ILDs are subject to the stress of traffic, they tend to fail quite frequently. Moreover, their installation and maintenance require lane closure and modifications to the pavement.

Microwave Radar Detectors

Microwave radar detectors are examples of nonintrusive detection devices whose installation and maintenance does not require lane closure and pavement modifications. As opposed to inductive loops, nonintrusive detection devices are not embedded in the pavement. Instead, they are typically mounted on a structure over or to the side of the road. Depending on the form of electromagnetic wave used, microwave radar detectors can measure either vehicle presence or vehicle presence and speed. One of the major advantages of microwave sensors stems from their ability to function under all weather conditions. Since these sensors are installed above the pavement surface, they are typically not subject to the effects of ice and plowing activities. Experience has shown that microwave sensors can be expected to function adequately under rain, fog, snow, or windy conditions.

Infrared Sensors

Infrared sensors are another example of nonintrusive devices. Infrared sensors are of two types: passive detectors and active detectors. Passive infrared detectors do not actually transmit energy (hence the name passive). Instead, they detect the energy that is emitted or reflected from vehicles, road surfaces, and other objects. Passive infrared detectors can measure speed, vehicle length, vehicle counts, and occupancy. The problem, however, is that their accuracy is affected by adverse weather conditions. Active infrared detectors, on the other hand, function in a manner somewhat analogous to microwave radar detectors by directing a narrow beam of energy toward the roadway surface. This beam is then directed back to the sensors, and vehicles are detected by noting changes in the round-trip propagation time of the infrared beam. Active infrared detectors supply vehicle passage, presence, speed, and vehicle classification information. Their accuracy, however, is easily affected by weather conditions such as fog and precipitation.

Ultrasonic Detectors

Ultrasonic vehicle detectors function in a manner analogous to microwave detectors by actively transmitting pressure waves at frequencies above the human audible range. These detectors can measure volume, occupancy, speed, and classification. Ultrasonic sensors, however, are sensitive to environmental conditions. Moreover, they require a high skill level to maintain.

Acoustic Detectors

Vehicular traffic produces acoustic energy or audible sound from a variety of sources within the vehicle and from the interaction of the vehicle's tires with the road surface. Using a system of microphones, acoustic detectors are designed to pick up these sounds from a focused area within a lane on a roadway. When a vehicle passes through the detection zone, a signal-processing algorithm detects an increase in sound energy, and a vehicle presence signal is generated. When the vehicle leaves the detection zone, the sound energy decreases below the detection threshold and the vehicle presence signal is terminated. Acoustic sensors can be used to measure speed, volume, occupancy, and presence. The advantages of acoustic sensors stem from their ability to function under all lighting conditions and during adverse weather.

VIP

VIP is a relatively recent traffic detection technique that promises to meet the future needs of traffic management and control. VIPs identify vehicles and their associated traffic flow parameters by analyzing imagery supplied by video cameras. The images are digitized and then passed through a series of algorithms that identify changes in the image background. A VIP system consists of a video camera, a digitizer, and a microprocessor for digitizing and processing the image, and software for interpreting the image and extracting detection information from it.

One big advantage of VIP systems stems from their ability to provide detection over a number of lanes and in multiple zones within the lane, thereby providing for true wide-area detection. In addition, the user can easily modify the detection zones, within seconds, through the graphical interface, without having to dig up the pavement or close traffic lanes. The performance of VIP systems, however, can be negatively affected by poor lighting, shadows, and inclement weather. Evaluation studies in Oakland County, Michigan, however, indicate that modern VIP systems yield excellent performance with a detection accuracy of more than 96% under all weather conditions [2].

VIP and CCTV

VIP systems can be combined with CCTV systems to provide an excellent detection tool, particularly for incident detection and verification purposes. When an incident occurs, the user can switch from the VIP mode to the standard CCTV mode and can then verify the occurrence of the incident via pan/tilt/zoom controls.

Vehicle Probes

The idea here is to use the vehicles themselves for detection by tracking them in time and space. This method has become feasible in recent years with the significant advancements in positioning and communications technologies. The location of the vehicle in time and space is communicated to a central computer, where data from different sources are fused to determine the status of traffic flow over the transportation system. Vehicle probes can actually provide for very useful information that other detection techniques cannot provide for, including information on link travel times, average speeds, and origin-destination information. Three different technologies that use vehicles as probes are currently available. These are *automatic vehicle identification* (AVI), *automatic vehicle location* (AVL), and anonymous mobile call sampling.

AVI technologies can be used to identify vehicles as they pass through a detection zone. Typically, a transponder (or a tag) mounted on the vehicle would be read by a roadside reader as the vehicle passes by at main line speeds.

This information can then be transmitted to a central computer. Currently, the most common application of AVI technologies in transportation is in conjunction with automatic toll collection systems (such as EZPass). With these systems, the value of the toll is automatically deducted from a driver's account each time the driver goes through the toll plaza. The same technology can also be used for detection purposes by determining the average travel time on freeways between roadside readers. This was actually done in Houston by using tag-equipped vehicles as probes to monitor traffic flow conditions.

AVL technologies allow for determining the location of vehicles in real time as they travel over the network. In addition to its use as a vehicle-probe detection method, AVL can be used for a host of other useful applications. For example, AVL can improve emergency management services by aiding in locating and dispatching emergency vehicles. AVL can be used on buses to help locate vehicles in real time and to determine their expected arrival time at bus stops. While a number of technologies are available for AVL systems, including dead reckoning, ground-based radio, signpost, and odometer, *global positioning systems* (GPS) are currently by far the most commonly used technology. For its operation, GPS relies on signals transmitted from 24 satellites orbiting the Earth at an altitude of 20,200 km. GPS receivers determine the location of a point by determining the time it takes for electromagnetic signals to travel from the satellites to the GPS receiver.

Anonymous mobile call sampling is based on using triangulation techniques to determine a vehicle's position by measuring signals resulting from the driver's use of a cellular phone within the vehicle. This concept promises to provide a wealth of information at a relatively low cost. This concept was first tested in the Washington, D.C., area in the mid-1990s. Currently, another test that uses anonymous mobile call sampling technology developed by U.S. Wireless is under way around the capital beltway area. The Universities of Maryland and Virginia have initiated a research project to evaluate the effectiveness of the approach [3].

Mobile Reports

Mobile reports constitute another very important source of freeway surveillance information that should not be overlooked. In many cases, reports of incidents from citizens and the police can provide significant system monitoring conditions at a very low cost compared to other surveillance technologies. Mobile reports do not provide a continuous stream of condition data like other surveillance technologies; rather, they provide event information at unpredictable intervals that could be very useful for traffic management purposes. In particular, mobile reports are very effective for incident detection. Examples of mobile reporting methods include cellular phones, call boxes, and freeway service patrols.

Cellular phones can serve as a very effective tool for incident detection. Many areas around the country have established an incident reporting hotline to encourage citizens' reporting of traffic incidents. This method has the advantage of low start-up costs. *Call boxes* are emergency phones that can be used for incident detection. Typically, the phones are located on the side of the freeway and are spaced at distances ranging from 0.25 to 0.50 mile apart. To use a call box, the motorist just needs to press a certain button to request the services of the police or emergency services provider. The widespread use of cell phones these days, however, can be considered a substitute for call boxes.

A *freeway service patrol* consists of a team of trained drivers who are responsible for covering a given segment of the freeway. The freeway service patrol vehicle would typically be equipped to allow for helping stranded motorists and clearing an incident site, if possible. Service patrols are capable of detecting incidents, as well as performing the entire incident management process (i.e., detecting and clearing an incident) in some cases.

Environmental Sensors

Environmental sensors can be used to detect adverse weather conditions such as icy or slippery conditions. This information can then be used to alert drivers via VMSs, for example; it can also be used by maintenance personnel to optimize maintenance operations. Environmental sensors can be divided into road condition sensors that measure surface temperature, surface moisture, and presence of snow accumulations; visibility sensors that detect fog, smog, heavy rain, and snow storms; and thermal mapping sensors that can detect the presence of ice. In addition, many manufacturers currently provide for complete weather stations that are capable of monitoring a wide range of environmental and surface conditions.

4.1.3.2 Hardware

Hardware constitutes the second basic component of a freeway surveillance system. Computers play a major role in the traffic surveillance process. They are responsible for the following:

- Receiving information from the field devices and sensors;

- Communicating data from the control center to the field devices (e.g., control data to provide for pan/tilt/zoom of a field CCTV camera);

- Data processing to derive meaningful traffic parameters from the real-time data collected by the sensors;

- Archiving the data collected to use for long-range planning, for example.

In addition to computer hardware, the surveillance system would typically include graphical displays in the control center to provide for a visual description of the transportation system operations as obtained from the field cameras. Graphics can be provided on the workstations' monitors or on large-screen graphics displays. These displays would typically take the form of an array of video screens.

4.1.3.3 Software

Computer software constitutes the third component of a traffic surveillance system. Without computer software, the data collected by the system's field devices are useless. Examples of software that a traffic surveillance system requires include incident detection algorithms, DSSs for incident management, and software to control field devices.

4.1.3.4 Communications System

The last component of a traffic surveillance system is the communications system, which is needed to provide communications between the different components within the control center, as well as between the center and the field devices. Communications within the center is typically accomplished via a *local-area network* (LAN). Communications between the center and the field devices, on the other hand, requires an external wireline system (i.e., fiber optic, coaxial cable, and twisted pair). Wireless communications could also be used in some cases. The choice of the communications medium would depend on the bandwidth requirements of the data transmitted. Video, for example, requires a wide bandwidth that can best be met using fiber-optic cables.

4.1.4 Ramp Control

As discussed in Chapter 2, ramp metering involves regulating vehicles' entry to the freeway system through the use of traffic signals, signs, and gates. Ramp metering aims to accomplish the following:

- To balance demand and capacity and to minimize operational breakdowns;
- To improve safety when some geometric deficiencies exist.

Ramp control systems currently operate in many areas around the country, including Minneapolis/St. Paul, Minnesota; Seattle, Washington; and Austin,

Texas. Most of these systems have achieved their goals in terms of reducing delay and improving safety. The operational concept of ramp metering is described in Chapter 2.

4.1.5 Incident Management

As previously mentioned, congestion on freeways is of two types, *recurrent* congestion and *nonrecurrent* congestion. Recurrent congestion, as the name implies, refers to a recurring congestion situation that occurs on a regular basis (for example, congestion during peak hours when the demand exceeds the capacity). Nonrecurrent congestion, on the other hand, is the result of unusual or abnormal occurrences, such as freeway accidents, stalled vehicles, work zone lane closures, or special events. The distinguishing feature between recurrent and nonrecurrent congestion is that while recurrent congestion is typically predictable, nonrecurrent congestion is not.

Incident management systems are mainly designed to deal with nonrecurrent congestion conditions. Essentially, incident management can be defined as "a coordinated and planned approach for restoring traffic to its normal operations after an incident has occurred." This incident could be a random event (such as a freeway accident or a stalled vehicle), and it could be a planned or scheduled event (such as work zone lane closures). In incident management, the goal is to *systematically* use both human and mechanical resources to achieve the following:

- Quickly detection and verification of the occurrence of an incident;

- Assessment of the severity of the situation and identification of the resources needed to deal with it;

- Determination of the most appropriate response plan aimed at quickly restoring the facility to normal operation.

The process of incident management can be conceptually viewed as consisting of the following four sequential stages:

1. Incident detection and verification;
2. Incident response;
3. Incident clearance;
4. Incident recovery.

The overall goal of the incident management process is to reduce the time needed for each of these four stages to allow for restoring normal operations in the most efficient fashion possible. A brief discussion of these four stages follows.

4.1.5.1 Incident Detection and Verification

Incident detection refers to the determination of the occurrence of an incident. Verification, on the other hand, refers to the determination of more details concerning the nature of the incident, such as its location, severity, and extent. Verification is crucial to providing the required information needed to devise an appropriate response plan. Before the advent of the ITS and advanced telecommunications era, incident detection and verification was primarily the responsibility of routine police patrols.

Technologies available for incident detection and verification can be divided into nonautomated and automated methods. Nonautomated detection techniques include cellular phone calls to 911 or to an incident reporting hotline, freeway service patrols, citizen-band radio monitoring, motorist call boxes, and fleet operators. These techniques often serve an important role in the incident management process, particularly when automated surveillance technologies are not available.

Automated surveillance technologies are among the most basic components of a freeway management system. These technologies can help dramatically improve the incident detection and verification process. *Automated incident detection* (AID) systems can be best regarded as consisting of two major components: a traffic *detection system* that provides the traffic information needed for detection, and an incident *detection algorithm* that interprets the information and determines whether an incident has occurred or not. Several AID algorithms have been proposed by researchers over the last two decades. These include comparative-type (California-logic) algorithms [4], the McMaster algorithm [5], and artificial intelligence-based techniques, including neural networks and fuzzy logic [6].

4.1.5.2 Incident Response

With an incident detected and verified, the next step in the incident management process is incident response, which involves activating, coordinating, and managing the personnel and equipment to clear the incident. Traffic incident response can be divided into two stages. Stage one is concerned with identifying the closest incident response agencies required to clear the incident, communicating with those agencies, coordinating their activities, and suggesting the required resources to deal with the incident effectively. The second stage involves the different traffic management and control activities aimed at reducing the adverse impacts of the incident. These include informing the public about the incident via VMSs and other information dissemination devices, implementing ramp metering and traffic diversion strategies, and coordinating corridorwide traffic control strategies.

The primary goal of incident response technologies is to optimize resource allocation and minimize response time, which can be divided into the following three components: the time needed to verify the occurrence and location of an incident; the dispatch time of the incident response team; and the travel time of the incident response time to the site. Reducing any of those three components will help reduce the overall response time. A number of techniques and technologies are available to help achieve this objective, as briefly described in the following [1].

Incident Response Manuals

Incident response manuals detail how to respond to specific types of incidents and who should be contacted. The manual also typically includes information on the locations and operators of large tow trucks and incident handling equipment. The goal of incident response manuals is to reduce response times by having adequate documentation describing how to deal with different types of incidents.

Incident Response DSSs

An incident response DSS typically takes the form of a knowledge-based system that can provide operators with useful information for developing response plans. A good example of these systems is the incident management real-time expert system developed by the University of California, Irvine, in Orange County [7]. The computer-based tool is designed to receive information from the user about the location and characteristics of the incident. The tool would then recommend an appropriate response and present that to the user in the order that the operator needs to follow.

Tow Truck Contracts

The idea here is to minimize response time by having preestablished contracts with towing companies. Tow trucks are typically called to an incident location by using a rotation list in which each wrecker service is called in order.

Techniques to Improve Emergency Vehicle Access and Traffic Flow

A number of techniques and technologies could be used to facilitate emergency vehicle access to the incident scene. These include the use of barrier openings and barrier gates, the use of emergency ramps, and shoulder utilization. Barrier openings can be used on freeways with inadequate access for emergency vehicles; however, they could encourage freeway motorists to use them for turning around if they miss an exit, which is not desirable from a safety standpoint. Barrier gates were introduced to address this problem. These gates remain closed until access is needed by an authorized vehicle. The use of shoulders close to incident locations has proven to be a successful strategy to improve emergency vehicle access and traffic flow, provided that the shoulder is wide enough. Shoulder utilization strategies are typically implemented by uniformed officers positioned upstream of the incident location.

Freeway Service Patrols

Dedicated freeway service patrols, which were previously discussed, can be very effective in reducing incident response time. As previously mentioned, a freeway service patrol consists of a team of trained drivers who are responsible for covering a given segment of the freeway. The freeway service patrol vehicle is equipped to help stranded motorists and clear some types of incident.

Traffic Management and Control

Many components of a freeway management system are designed to assist in incident management. These components include information dissemination devices (such as dynamic message signs and highway advisory radio) that could be used to warn drivers of downstream incidents and traffic conditions. They also include ramp metering components that could be used to reduce traffic demands entering the freeway in the vicinity of an incident.

Alternative Route Planning

Alternative route planning involves whether diverting traffic should be implemented or not, determining how much traffic should be diverted off the freeway when an incident occurs, and determining the alternative routes to which this traffic should be diverted. Diverting traffic to alternate routes is often politically sensitive, and typically the severity and the likely duration of an incident would play a major role in determining whether diversion should be implemented or not. In the past, alternative route plans were typically prepared on hard-copy printouts. More recently, however, these plans are being converted to a digital format on *geographic information systems* (GIS) to assist in organization and retrieval.

4.1.5.3 Incident Clearance

Incident clearance refers to the safe and timely removal of an incident. While around 80% of the incidents in urban areas are minor incidents that do not typically require a towing truck, the remaining 20% do. When a towing truck is needed, the police officer at the site would typically be the one responsible for overseeing the safety of the operation. There are several technologies for improving the efficiency of incident clearance, including inflatable air bag systems. The main purpose of these systems is to right overturned vehicles. Typically, the system consists of rubber inflatable cylinders having various heights. These cylinders are placed underneath the overturned vehicle and inflated to right the vehicle.

4.1.5.4 Incident Recovery

This stage refers to the time taken by traffic to return to normal after the incident has been cleared. The goal here is to use sound traffic management techniques to restore normal operations and to prevent the effect of congestion from spreading elsewhere.

4.1.6 Information Dissemination

Effective communication with drivers is an essential component of the freeway management process. Freeway management systems use several traffic information dissemination devices to keep drivers informed about current as well as expected travel conditions on the freeway. These devices include VMSs, highway advisory radio, cable TV, kiosks, and others. Typically, traffic information dissemination is divided into pretrip information dissemination and en route information dissemination.

Pretrip traffic information concerns providing travelers with information before they start their trips. Examples of pretrip traveler information includes current or expected traffic conditions, current and expected weather conditions, and bus schedules and fares; this information is intended to enable travelers to make informed route, mode, and time-of-departure decisions. En route traffic information dissemination, on the other hand, involves providing travelers with information while they are en route. En route traffic information includes many of the same items provided for pretrip planning, such as current and expected traffic and weather conditions, traffic incidents, and suggested diversion routes.

The basic goal of information dissemination is to provide travelers with real-time information on the status of the transportation system to allow travelers to make more informed decisions regarding the time of departure for the trip, the transportation mode to be utilized, and the specific route to be taken. The implementation of the information dissemination component of a freeway management system would typically require cooperation among a number of partners, including transportation and public works departments, transit agencies, toll authorities, law enforcement, commercial media, and private-sector traffic reporting services.

The effectiveness of information dissemination is a function of at least three factors: the credibility of the information provided; market penetration; and traveler response. For the information provided to have a positive impact on traffic operations, it has to be credible. This means that the information provided must be timely, accurate, and relevant to its intended audience. Market penetration, which refers to the percentage of the potential audience reached by information dissemination efforts, also impacts the effectiveness of an information dissemination system. Finally, the effectiveness will depend on travelers' response to the information being provided. Information dissemination systems can be used to perform a number of functions. These include notifying motorists of downstream conditions, advising motorists when to seek alternate routes, waning motorists of adverse weather conditions, and informing drivers of construction work activities, among others.

The different information dissemination devices can be divided into three groups: on-roadway information devices; in-vehicle information devices; and off-roadway information for pretrip information dissemination. A brief discussion of these three groups follows.

4.1.6.1 On-Roadway Information Devices

The function of on-roadway information devices is to provide drivers with information while they are en route. *Dynamic message signs* (DMS) are among the most common types of devices used for this purpose. DMS can either be fixed or portable.

DMS can be used to provide travelers with useful information on traffic and weather conditions, incident locations and expected delays, construction work activities, suggested alternate routes, and speed advisories, among others. DMS can be divided, based on the technology they use, into light-reflecting, light-emitting, and hybrid [1].

- *Light-reflecting signs* reflect light from an external source (such as the Sun, headlights, and overhead lighting). Among the different types of light-reflecting signs are rotating drum and reflective disk matrix signs. Rotating drum signs are made of one to four multifaced drums, each containing two to six messages. Reflective disk matrix signs, on the other hand, are comprised of an array of permanently magnetized, pivoted indicators that are black on one side and reflective yellow on the other. When a given pixel is to be activated, an electric current is used to flip the indicator from the black finish to the reflective yellow finish. Reflective disk matrix signs were quite popular in the 1970s for freeway management systems because they are less costly than light-emitting signs.

- *Light-emitting signs*, as opposed to light-reflecting signs, generate their own light on or behind the viewing surface. *Light-emitting diode* (LED) and fiber-optic DMS are two examples of light-emitting signs. LED DMS use LED to create display pixels. LEDs are typically amber in color. Fiber-optic DMS funnel light energy from a light source through fiber bundles to the sign face.

- *Hybrid DMS* combine the characteristics of both light-reflecting and light-emitting DMS. One of the best examples of hybrid DMS is the reflective disks and fiber optic or LED DMS. During weather conditions when light-reflecting DMS are not clearly visible, these hybrid systems would use light-emitting technologies such as fiber optics or LED. When the sun is out, the light sources are turned off.

4.1.6.2 In-Vehicle Information Devices

These devices are typically located inside the vehicle and, similar to DMS, are designed to provide information to drivers while en route. In-vehicle information devices can provide information by either audio or visual means. Examples of auditory in-vehicle information devices include *highway advisory radio* (HAR), cellular phone hotlines, and commercial radio. Examples of visual in-vehicle devices include video display devices and head-up displays.

HAR

HAR provides another means for disseminating information to drivers while en route. Typically, information is provided through the AM receiver. To inform drivers about the existence of an HAR signal, signs are typically installed upstream of the signal, instructing drivers to tune to a given frequency (typically either 530 or 1,610 kHz). HAR can be used to provide travelers with information similar to that provided by DMS. One advantage of HAR compared to DMS, however, stems from the fact that it is less distracting, since information is provided through a different sensory channel (audio), which helps reduce information overload received visually. In addition, more complex messages are possible with HAR compared to DMS. The disadvantage, however, is that people have to tune to the frequency themselves.

Cellular Telephone Hot Lines

Another way to provide information to drivers en route, which has increased in popularity with the widespread use of cellular phones, involves establishing a special hot line phone system for traffic information that drivers can call from their cell phones while en route. Examples of these systems include the Smar-Traveler System in Boston. The phone systems would typically include a touch-tone menu that allows callers to receive route-specific traffic information; this gives drivers control over the type of information they receive. A national effort is currently under way to establish a national phone system that people can access to get traveler information. This effort will designate 511 as a national traveler information system.

Commercial Radio

Commercial radio is yet another means of providing en route traveler information. The primary disadvantage of commercial radio, however, pertains to the accuracy and timeliness of the information. Typically, information is broadcast when normal scheduling permits; in many cases, this may not be inappropriate since an incident might have already been cleared by the time normal scheduling permits broadcasting.

Video and Head-Up Display Terminals

A recent approach to disseminating traffic information en route involves the use of video and head-up display terminals.

4.1.6.3 Off-Roadway Information

In addition to the aforementioned en route traffic information dissemination devices, several other devices could be used for pretrip, off-roadway information dissemination. These include cable TV, phone systems, the Internet, pagers, *personal digital assistants* (PDAs), and kiosks. For example, many metropolitan areas around the country now have Web sites dedicated to traveler information. Examples of these roadway information sites include the Houston TranStar system developed by the Texas Department of Transportation; CHART, developed by the Maryland State Highway Administration; and the Virginia Department of Transportation Web site. These systems provide travelers with a wealth of travel-related information, including current travel conditions, alerts, and other timely information. A traffic map showing current speeds, locations of any incidents or construction zones typically form the central piece of these Web sites.

4.1.7 Real-World Freeway and Incident Management Systems and Their Benefits

Real-world freeway and incident management systems can now be found throughout the country. Examples include systems in Atlanta, Houston, Seattle, Minneapolis/St. Paul, New York, Chicago, Milwaukee, Los Angeles, San Diego, and Northern Virginia, among others. Freeway and incident management has been proven to be quite effective in alleviating recurrent and nonrecurrent congestion. The San Antonio's TransGuide freeway management system in Texas, for example, helped reduce accidents by 15% and emergency response time by 20%. Ramp metering was shown to help increase throughput by 30% in the Minneapolis/St. Paul metro area, with peak-hour speeds increasing by 60%. Ramp meters in Seattle are credited with a 52% decrease in travel time and 39% decrease in accidents [8]. The evaluation of the initial operation of the Maryland CHART program showed a benefit/cost ratio of 5.6:1, with most of the benefits resulting from a 5% decrease (which amounted to around 2 million vehicle-hours per year) in delay from nonrecurrent congestion.

4.2 Advanced Arterial Traffic Control Systems

Advanced arterial traffic control systems (or adaptive traffic control systems) are adaptive control systems that depend on the use of a digital computer and sophisticated computer algorithms to control the operation of a system of signals along an arterial. The basic idea is to take advantage of the power of digital computers to control many signals, along an arterial or in a network, from one central location. Computer traffic control systems predate ITS by at least several decades; the first installation of these systems took place in the early 1960s. These systems, however, have undergone continuous refinement since that time. The following provides some historical perspective on the development of these systems.

4.2.1 Historical Development of Computer Traffic Control Systems

4.2.1.1 Systems with No Feedback

The most basic type of computer signal control systems first appeared in the 1960s. The idea is for a computer to control a series of controllers, but with no "feedback" of information from the field detectors back to the computers (open-loop control). In such a system, the traffic plans implemented are not responsive to the actual traffic demand. Instead, the plans are developed "off-line" from historical traffic counts and implemented based on the time of the day and the day of the week. While this system appears rather simplistic, it still offers the following advantages:

- *The ability to update signal plans from a central location greatly facilitates the implementation of new plans as the need arises.* Experience clearly shows that in the majority of major cities in the United States, signal plans are not updated frequently enough, largely because of the rather time-consuming task of sending signal technicians to the field to retime the controllers. A computer-based system eliminates this problem.

- *A computer signal control system allows for the storage of a large number of signal plans that could be implemented based on the prevailing traffic conditions.* For example, one could have a different plan for weekends, a different plan for special events such as a college football game or a county fair, and a different plan for snow events.

- *A computer signal control system also allows for the automatic detection of any malfunction in the operation of the controllers or the signal heads.* This allows for the timely dispatch of maintenance crews.

4.2.1.2 Systems with Feedback

The next development was to have signal control systems where information from the field traffic detectors is fed back into the central computer. The central computer would then use this information to select the signal plan to be implemented (closed-loop control). Plan selection is conducted according to one of the following methods.

1. *Select plan from a library of predeveloped plans.* In this method, the system has access to a database (library) that stores a large number of different traffic patterns along with the "optimal" signal plans for each pattern (these plans are developed off-line). Based on information from the traffic detector, the computer matches the observed traffic pattern against the patterns stored in the library and identifies the closest stored pattern. The plan associated with the identified pattern is then implemented.

    This type of adaptive traffic control system is often referred to as a *first generation* system. The distinguishing feature of these systems is that the plans, while responsive to traffic conditions, are still developed off-line. Typically, the frequency of signal update is every 15 minutes. First generation systems do not generally have traffic prediction capabilities.

2. *Develop plan on-line.* In this method, the "optimal" signal plan is computed and implemented in real time. This obviously requires enough computational power to do the necessary computations on-line. Systems that develop plans on-line are classified as either *second generation* or *third generation* systems. These systems typically have a much shorter plan update frequency compared to *first generation* systems. In addition, signal plans are computed in real time based on forecasts of traffic conditions obtained from feeding the detectors information into a short-term traffic-forecasting algorithm. For second generation systems, the plan update frequency is every 5 minutes, whereas third generation systems have an update interval ranging from 3 to 5 minutes. In Section 4.2.2, we will describe some examples of these adaptive traffic control systems that are in use in the United States and around the world.

4.2.2 Adaptive Traffic Control Algorithms

A number of adaptive traffic control algorithms are currently available. Among the most widely accepted are SCOOT, SCATS, and RHODES. A brief description of these algorithms follows.

4.2.2.1 SCOOT

SCOOT (which stands for Split, Cycle, Offset Optimization Technique) is an adaptive traffic control system developed by the *U.K. Transport Research Laboratory* (TRL) in the early 1980s. In 1996, the system was in operation in more than 130 towns and cities all around the world. SCOOT operates by attempting to minimize a *performance index* (PI), which is typically the sum of the average queue length and the number of stops at all approaches in the network. To do this, SCOOT modifies the cycle lengths, offsets, and splits at each signal in *real time* in response to the information provided by the vehicle detectors.

SCOOT can be regarded as an on-line version of the famous TRANSYT-7F model described in Chapter 2, since it essentially uses the same methods that TRANSYT-7F employs [9]. However, instead of just having fixed timed plans that age quickly as a result of volume changes, SCOOT computes and implements the "best" timing plan automatically in real time in response to varying traffic conditions.

The operation of SCOOT is based on *cyclic flow profiles* (CFPs), which are histograms of traffic flow variation over a cycle, measured by loops and detectors and placed midblock on every significant link in the network. Using the CFPs, the offset optimizer calculates the queues at the stop line. The optimal splits and cycle length are then computed.

In recent years, a number of features have been added to SCOOT to improve its effectiveness and flexibility. These include the following:

- The ability to provide preferential treatment or signal priority for transit vehicles;

- The ability to automatically detect the occurrence of incidents;

- The addition of an automatic traffic information database that feeds historical data into SCOOT, allowing the model to run even if there are faulty detectors [10].

4.2.2.2 SCATS

The *Sydney Coordinated Adaptive Traffic System* (SCATS) was developed in the late 1970s by the Roads and Traffic Authority of New South Wales in Australia. To operate, SCATS requires only stop-line traffic and not midblock traffic detection, as does SCOOT. This is definitely an advantage, since the majority of existing signal systems are equipped with sensors only at stop lines. SCATS is a distributed intelligence, hierarchical system that optimizes cycle length, phase intervals (splits), and offsets in response to detected volumes. For control, the whole signal system is divided into a large number of smaller subsystems ranging from one to 10 intersections each. The subsystems run individually unless traffic conditions require the "marriage" or integration of individual subsystems.

In developing the real-time signal plans, SCATS' objective is generally to equalize the saturation flow ratio of conflicting approaches. Consequently, the system in many cases does not minimize delays on major arterials, which may actually exhibit deterioration in their level of service, particularly during peak periods. This was quite evident in the FAST-TRAC ITS field test in Oakland County, Michigan. In that project, video detection was used to feed a SCATS system, which then developed timing plans in real time.

4.2.2.3 RHODES

Since 1991, the University of Arizona has been developing a real-time adaptive control system called RHODES, which stands for Real-time Hierarchical Distributed Effective System [11]. RHODES is designed to take advantage of the natural stochastic variations in traffic flow to improve performance, a feature that is missing from other systems such as SCOOT and SCATS. The system consists of a three-level hierarchy that decomposes the traffic control problem into three components: network loading, network flow control, and intersection control (Figure 4.3).

At the highest level, a stochastic traffic equilibrium model is used to predict the expected traffic loads on the links of the network. The prediction time horizon for level 1 is typically in the order of 15-minute intervals. The second level, level 2, represents the high-level decision-making process for setting signal timing to optimize traffic flow. This level recognizes the stochastic nature of traffic and attempts to take into account future expected traffic loads over the next few minutes. Level 2 is concerned with setting approximate phase sequences and splits for a given corridor (target timings). Finally, level 3 is concerned with intersection control—determining the optimal light change epochs for the next phase sequence and determining whether the current phase should be shortened or extended. The time frame for control level 3 is typically in the order of seconds and minutes. At the time of this writing, preparations are under way for a field test of RHODES. It is envisioned and hoped that RHODES will produce better results than SCOOT and SCATS.

4.2.3 Real-World Adaptive Traffic Control Systems and Their Benefits

Adaptive traffic control systems offer a higher degree of control responsiveness to traffic conditions, and therefore, *if applied in the right context,* should be expected to result in benefits that are similar yet somewhat higher than those discussed in relation to signal coordination projects. Some examples of real-world systems, along with their reported benefits, are briefly listed next [8].

ITS Applications and Their Benefits

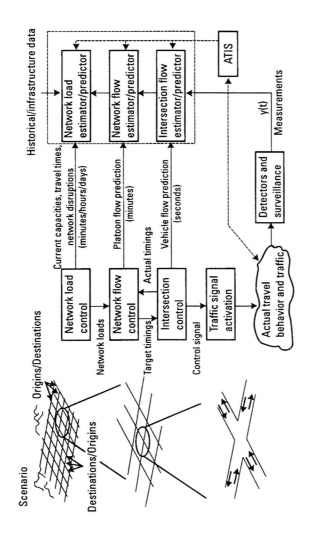

Figure 4.3 RHODES three-level hierarchical architecture. (*From*: [11]. © 1992 Transportation Research Record.)

Oakland County, Michigan

As a part of the FAST-TRAC program, the SCATS for signal control became operational in Troy, Michigan, in 1992. Evaluation studies of the systems' effectiveness showed a decrease of 33% in the number of stops, as well as increased average speeds, particularly during off-peak periods. From a customer's satisfaction standpoint, 72% of drivers said they are better off with FAST-TRAC.

Los Angeles, California

A computerized system has been in operation since 1984. As of 1994, the system includes 1,170 intersections and 4,509 detectors for signal timing optimization. It has been reported that the system resulted in a 13% decrease in fuel consumption, 14% decrease in emissions, 41% reduction in vehicle stops, 18% reduction in travel time, 16% increase in average speed, and 44% decrease in delay.

Toronto, Ontario

In Toronto, the SCOOT adaptive traffic signal control system was used to control 75 signals within the metropolitan area. When compared to a best-effort fixed timing plan, the evaluation showed an 8% decrease in travel time, 22% decrease in vehicle stops, and 17% decrease in vehicle delay.

4.3 Advanced Public Transportation Systems

Advanced public transportation systems attempt to improve the efficiency, productivity, and safety of transit systems. They also strive to increase ridership levels and customer satisfaction. In this section, we describe some examples of advanced public transportation systems. These examples can be categorized under the following categories: AVL systems; transit operations software; transit information; and electronic fare payment systems.

4.3.1 AVL Systems

AVL systems are designed to allow for tracking the location of transit systems in real time. These systems work by measuring the actual real-time position of each vehicle and communicating the information to a central location. This information can then be used to increase dispatching and operating efficiency, allow for quicker response to service disruptions, provide input to transit information systems, and increase driver and passenger safety and security.

While a number of technologies are available for AVL systems, including dead reckoning, ground-based radio, signpost and odometer, and GPS, most agencies are choosing GPS-based systems. GPS is a navigational and positioning system that relies on signals transmitted from satellites for its operation. In

1996, 86 transit agencies across the country operated, implemented or planned AVL systems, more than 80% of which used GPS technologies. Some real-world implementations of AVL systems are described next [12].

Atlanta, Georgia

About 250 buses of the *Metropolitan Atlanta Rapid Transit Authority* (MARTA)'s 750-vehicle fleet are equipped with AVL. The system is linked to the Georgia Department of Transportation's traffic management center. There are also electronic signs at a few bus stops to display information to waiting passengers. Although real-time bus location information is not yet fed into the many electronic passenger information kiosks around the city, there are plans to do so in the future. The system has shown to yield concrete benefits, including the ability to more effectively improve on-time performance as well as increase safety.

Portland, Oregon

The Tri-County Metropolitan Transportation District of Oregon (Tri-Met) has recently completed the implementation of a GPS AVL system for 640 fixed-route vehicles and 140 paratransit vehicles. The AVL is being employed as part of a regional ITS system, whereby the buses will be used as probes for traffic monitoring, as previously discussed.

Denver, Colorado

The *Regional Transportation District* (RTD) has had an operational AVL system on all of its 900 buses since 1995. AVL data are used to post real-time departure information on signs at the two mall stations downtown. The system also includes an extensive computer-aided dispatch system. Current plans are to put passenger information data on the Internet and to feed it to information kiosks around the city.

Milwaukee, Wisconsin

The *Milwaukee Transit System* (MTS) completed the installation of a GPS AVL on 543 buses and 60 support vehicles. Preliminary results indicate a 28% decrease in the number of buses more than 1 minute behind schedule.

Ann Arbor, Michigan

The *Ann Arbor Transportation Authority* (AATA) has installed an AVL system on its entire fleet, consisting of 70 fixed-route buses and 10 paratransit vehicles. At dispatch, there are computer stations controlling fixed-route and paratransit operations. The AVL feeds an extensive pretrip and wayside real-time passenger information system.

4.3.2 Transit Operations Software

Transit operations software allow for automating, streamlining, and integrating many transit functions. This includes computer applications such as *computer-aided dispatching* (CAD), service monitoring, supervisory control, and data acquisition. The use of operations software can improve the effectiveness of operations dispatching, scheduling, planning, customer service, and other agency functions. Operations software is available for fixed-route bus operations, as well as for paratransit or demand-responsive operations. The following describes some successful real-world implementations of such systems and their benefits [12].

4.3.2.1 Fixed-Route Bus Operations

Kansas City, Missouri, was able to reduce up to 10% of the equipment required for bus routes using an AVL/CAD system. This allowed Kansas City to recover its investment in the system in two years. On-time performance improved by 12% in the first year of operating the AVL system.

The bus system in Sweetwater County, Wyoming, has *doubled* its monthly ridership using a CAD system; 5 years after its installation, transit-operating costs have decreased 50%.

4.3.2.2 Demand Responsive Transit Operations

Operations software for demand response transit implement new scheduling and dispatching software for improved performance and increased passenger-carrying capability of the vehicles. Systems vary widely in their capabilities; the high-end systems have integrated automated scheduling and dispatching software with AVL, GIS, and advanced communications systems. These systems provide dispatchers with the capability to view maps of the service area with the locations of all the vehicles in real time. Drivers have mobile data terminals displaying the next hour's pickups and drop-offs. It should be noted, however, that although only a few agencies have installed the most sophisticated of these systems, many have implemented one or more of their features. Some examples of such systems follow [12].

Ann Arbor, Michigan

Ann Arbor's paratransit service (A-Ride) has implemented CAD, automated scheduling, and advanced communications for eight AVL-equipped paratransit vehicles. This system is able to provide service 24 hours per day, with a dispatcher needed only to take reservations and cancellations from callers and to confirm rides.

Houston, Texas

METROLift installed AVL on all of its 110 vans and 55 sedans in 1993 and has recently implemented scheduling and dispatching software to be used in conjunction with the AVL for real-time scheduling. Evaluation results show an improvement in several performance measures, including location accuracy, service efficiency, and customer and dispatcher satisfaction.

4.3.3 Transit Information Systems

Transit information systems implement traveler information systems that provide travelers with transit-related information. Three types of systems can be identified: pretrip systems; in-terminal/wayside; and in-vehicle. A brief description, along with real-world examples, of each of these types follows.

4.3.3.1 Pretrip Transit Information Systems

These systems provide travelers with accurate and timely information before starting their trips to allow them to make informed decisions about modes, as well as routes and departure times. Pretrip information can cover a range of categories, including transit routes, maps, schedules, fares, park-and-ride locations, points of interest, and weather. In addition, they often support itinerary planning. Methods of obtaining pretrip information include touch-tone phones, pagers, kiosks, the Internet, fax machines, and cable TV. A few examples of such systems are given next.

Seattle, Washington

A key product of Seattle Metro's transit information system is a Web site where travelers can obtain information on transit schedules and fares, van and carpooling, ferries, and park-and-ride facilities. This site also provides assistance to transit users in planning their trips. In addition, the University of Washington has developed a Java applet to allow users to view the locations of all buses traveling throughout the Metro system (see Figure 4.4) [13]. The university has also developed Web pages to help travelers predict the time of arrival of buses at different bus stops (see http://mybus.org).

Winston-Salem, North Carolina

As a part of the second phase of its Advanced Public Transportation Systems Mobility demonstration project, the Winston-Salem Transit Authority plans to integrate fixed-route trip planning software into its existing CAD and scheduling software system, install an integrated and automated telephone system to provide touch-tone user information, and provide real-time paratransit and transit information via telephone, cable TV, and kiosks.

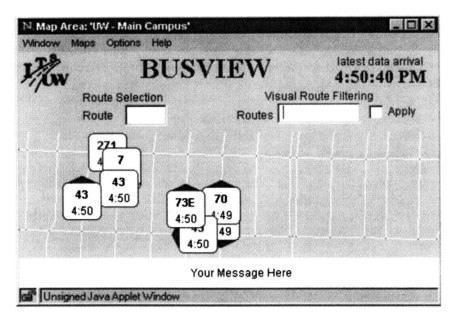

Figure 4.4 Busview Java applet showing real-time bus locations. (*From:* [14]. © 2002 Intelligent Transportation Systems University of Washington.)

Vancouver, British Columbia

BC Transit in Vancouver makes trip itinerary planning available to its customers via the telephone. A clerk answers the calls and inputs the origin and destinations and other information into the system, which is linked to a GIS map display showing the roads, bus stops, and significant points of interest. The software produces two or three optional itineraries, and the clerk informs the caller of the results.

4.3.3.2 In-Terminal/Wayside Transit Information Systems

These systems provide information to transit riders who are already en route. This information is typically communicated using electronic signs, interactive information kiosks, and CCTV monitors. The overall goal is to provide real-time bus and train arrival and departure times, reduce waiting anxiety, and increase customer satisfaction.

With respect to interactive information kiosks, the experiences of agencies that have already installed kiosks provide some useful information. The experience of Seattle Metro, for example, clearly points to the importance of the location and content. Ann Arbor Transportation Authority's research indicates that a Web site should be in place first and that kiosks should have the same interface as the Web site. Moreover, it found that itinerary planning applications using real-time information is key to successful deployment. Finally, several projects are attempting to secure private-sector partners to make these systems financially self-sustaining. The following are some examples of such systems.

Cincinnati, Ohio

The *Southwest Ohio Regional Transit Authority* (SORTA) is developing the infrastructure needed to deploy in-terminal information kiosks and onboard annunciators and displays. The infrastructure's components include a new 10-channel radio system and installation of vehicle logic to support AVL, CAD, and schedule-adherence systems.

Ann Arbor, Michigan

Using a pair of 31-inch video monitors, Ann Arbor Transit Authority will display the real-time data generated by the AVL system to inform passengers at its downtown transit center about arrival status, delays, and departure times. The operations center will also be able to display real-time incident information that will assist passengers in making alternate plans if their bus is delayed.

Phoenix, Arizona

As part of its AZTech Model Deployment Initiative, Phoenix will equip 88 buses with an AVL system. The AZTech server will process all real-time and static transit and traffic information for distribution through the system's three information channels, including feeding displays at 10 or more bus stops with real-time, next-bus arrival information. Five interactive kiosks—one at the downtown Phoenix transit terminal, one at the North Valley terminal, and three at malls served by major routes—will also be deployed.

4.3.3.3 In-Vehicle Transit Information Systems

The major impetus for in-vehicle customer information systems is to provide useful en route information for travelers onboard the vehicle and to comply with the applicable provisions of the American Disability Act of 1991. A good example is Ann Arbor's system, which includes in-vehicle annunciators and displays from which passengers will receive next-stop and transfer information. The latter will take the form of announcements that identify valid bus transfers at upcoming stops.

    A special feature of the Ann Arbor system is the ability for bus operators to communicate transfer requests. When a bus operator receives a transfer request from a passenger, he or she inputs the information into the onboard console, which allows the request to be processed by the dispatch computer. The dispatch computer then calculates whether the transfer can be made given the schedule adherence of both the originating and connecting buses. If the transfer is possible, the bus driver is informed to notify the requesting passenger, and the operator on the connecting bus is automatically advised to wait for a specified duration for the transfer.

4.3.4 Electronic Fare Payment Systems

The basic idea behind electronic fare payment systems is to facilitate the collection and management of transit fare payments by using electronic media rather than cash or paper transfers. Basically, these systems consist of two main components: a card and a card reader. Cards could be of the magnetic-stripe type, where the reader does most of the data processing. They could also be equipped with a microprocessor (smart cards), in which case the data processing could occur on the card itself. Electronic fare payment systems offer the following advantages:

- Eliminates the need for any actions to be taken on the part of vehicle operations;
- Eliminates the need for a passenger to worry about having the exact change for the bus fare;
- Facilitates fare collection and processing;
- Allows for more complex and effective fare structures to be adopted.

There are basically two types of electronic fare payment systems: closed systems and open systems. Closed systems are limited to one main purpose (i.e., paying transit fares) or to a few other applications, such as paying parking fees. However, the value stored on the card cannot be used outside the defined set of activities, hence the "closed system" name. Open systems, on the other hand, can be used outside the transit system. A prime example of an open system is a credit card, which naturally can be used with multiple merchants.

4.4 Multimodal Traveler Information Systems

Multimodal traveler information systems are designed to provide static, as well as real-time, travel information over a variety of transportation modes (e.g., highways, transit, ferries, and so on). In essence, these systems integrate the traffic information dissemination functions of freeway and incident management systems (see Section 4.1.6) with the functions of transit information systems (see Section 4.3.3). They then add more information from sources such as the Yellow Pages, tourist organizations, and weather services. As previously discussed, traveler information can be provided before or during a trip (pretrip and en route traveler information). Table 4.1 shows an extensive list of data of potential interest to the traveler that could be part of the traveler information system [15]. As can be seen, information is classified as either static or real time.

Following the collection and processing of the data, telecommunications technologies, including voice, data, and video transmission over wireline and wireless channels, are then used to disseminate the information to the public. Among the means of information dissemination are the Internet, cable TV, radio, phone systems, kiosks, pagers, PDAs, and in-vehicle display devices. Multimodal traveler information services allow users to make more informed decisions regarding time of departure, routes, and mode of travel. They have been shown to increase transit usage as well as reduce congestion when travelers choose to defer or postpone trips or select alternate routes.

A good example of a regional multimodal traveler information system is the Smart Trek Model Deployment Initiative in Seattle. At the center of this system is a set of protocols and paradigms designed to collect and fuse data from a variety of sources, process the data to derive useful information, disseminate the information derived to independent information service providers, and warehouse the data for long-range planning. Figure 4.5 shows a Web capture of

Table 4.1
Potential Contents of a Multimodal Traveler Information System

Static information—known in advance, changes infrequently	Planned construction and maintenance activities
	Special events, such as state fairs and sporting events
	Transit fares, schedules, and routes
	Intermodal connections (e.g., ferry schedules along Lake Champlain)
	Commercial vehicle regulations (such as hazardous material and height and weight restrictions)
	Parking locations and costs
	Business listings, such as hotels and gas stations
	Tourist destinations
	Navigational instructions
Real-time information—changes frequently	Roadway conditions, including congestion and incident information
	Alternate routes
	Road weather conditions, such as snow and fog
	Transit schedule adherence
	Travel time

Source: [15].

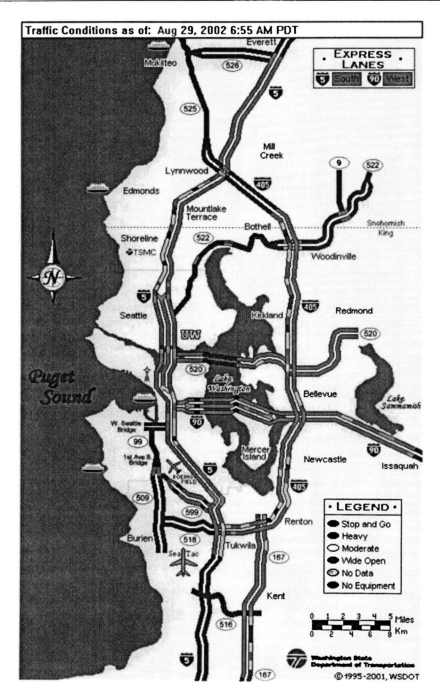

Figure 4.5 Smart Trek freeway map in Seattle. (*From:* [16]. © 2002 Washington State Department of Transportation. Used with permission.)

the Smart Trek Web page. In addition to providing speed and congestion information, the Web site is designed to provide travelers with information on bus arrival times, ferry schedules, weather conditions, and incident reports. The Web page is also used to support a TV channel, which was initially deployed on the University of Washington's cable station in 1998.

Benefits of Multimodal Traveler Information Systems

Although multimodal traveler information systems have the potential to yield significant benefits for both travelers and system operators, the demand for products has been slow to materialize. The size of the market for traveler information systems has been modest to date. Several reasons could be provided for the slow growth of the traveler information market. First, consumer awareness of traveler information products is currently quite low. Second, the price of some products, especially in-vehicle display devices, is still high. Finally, the quality of the information and the extent of coverage need to be increased [17].

4.5 Conclusions

In this chapter, we looked at some examples of real-world ITS applications that would typically be found in a metropolitan area. The examples discussed can be broadly categorized into four applications areas, namely freeway and incident management systems, advanced arterial traffic control systems, advanced public transportation systems, and multimodal traveler information systems. For each application, the operational concept was first discussed, followed by some examples of real-world deployments and their benefits. ITS has the potential to solve some of our nation's most challenging transportation problems. In the following chapters, we will discuss the process for determining the need for ITS deployment in a particular region and describe how to develop an ITS strategic plan to guide its deployment in a fashion that would ensure that it is deployed in response to identified problems in the region's transportation system.

Review Questions

1. What is the difference between recurrent and nonrecurrent congestion?
2. List the primary objectives of an FIMS.
3. Discuss the different functions an FIMS should be designed to perform.
4. Describe the four basic components of a traffic surveillance system.

5. Describe five different detection methods. Discuss the advantages and disadvantages of each.
6. What is the basic idea behind using vehicle probes as a detection method? Briefly describe some of the different technologies that use vehicles as probes.
7. Briefly discuss the four different stages of the incident management process.
8. Discuss some of the benefits reported for FIMS.
9. What are the three levels of control in the RHODES signal adaptive traffic control system?
10. Give some examples for real-world transit tracking systems and transit information systems.
11. Give some examples for both static as well as dynamic data elements that would be included in a multimodal traveler information system.
12. What are some of the functions performed by the Smart Trek multimodal traveler information system?

References

[1] U.S. Department of Transportation, Federal Highway Administration, *Freeway Management Handbook*, Report No. FHWA-SA-97-064, 1997.

[2] Michalopoulos, P. G., et al., "Field Deployment of Machine Vision in the Oakland County ATMS/ATIS Project," *Proc. IVHS America 1994 Annual Meeting*, Atlanta, GA, April 1994, pp. 335–342.

[3] Smith, B. L., et al., "Transportation Management Applications of Anonymous Mobile Call Sampling, *Proc. 11th Annual Meeting of ITS America*, Miami, FL, 2001.

[4] Payne, H. J., and S. C. Tignor, "Freeway Incident Detection Algorithms Based on Decision Trees with States," *Transportation Research Record 682*, TRB, National Research Council, Washington, D.C., 1978, pp. 30–37.

[5] Persaud, B., and F. L. Hall, "Catastrophe Theory and Pattern in 30-Second Freeway Traffic Data—Implication for Incident Detection," *Transportation Research A*, Vol. 23, No. 2, 1989, pp. 103–113.

[6] Cheu, R. L., and S. G. Ritchie, "Automated Detection of Lane-Blocking Freeway Incidents Using Artificial Neural Networks," *Transportation Research C*, Vol. 3, No. 6, 1995, pp. 371–388.

[7] Ritchie, S. G., and N. A. Prosser, "Real-Time Expert System Approach to Freeway Incident Management," *Transportation Research Record 1320*, TRB, National Research Council, Washington, D.C., 1991, pp. 7–16.

[8] Mitretek Systems, Inc., U.S. Department of Transportation, Federal Highway Administration, *ITS Benefits: 1999 Update*, Report No. FHWA-OP-99-012, 1999.

[9] Hansen, B. G., P. T. Martin, and H. J. Perrin, Jr., "SCOOT Real-Time Adaptive Control in a CORSIM Simulation Environment," *Transportation Research Record 1727*, TRB, National Research Council, Washington, D.C., 2000.

[10] Bretherton, D., "Current Developments in SCOOT: Version 3," *Transportation Research Record 1554*, TRB, National Research Council, Washington, D.C., 1996.

[11] Head, K. L., P. B. Mirchandani, and D. Sheppard, "Hierarchical Framework for Real-Time Traffic Control," *Transportation Research Record 1360*, TRB, National Research Council, Washington, D.C., 1992.

[12] U.S. Department of Transportation, Federal Transit Administration, *Advanced Public Transportation Systems: The State of the Art—1998 Update*, Report No. FTA-MA-26-7007-98-1, 1998.

[13] Dailey, D. J., *Smart Trek: A Model Deployment Initiative*, U.S. Department of Transportation, 2001, http://www.its.washington.edu/pubs/smart_trek_report.pdf.

[14] http://www.its.washington.edu/projects/busview_overview.html.

[15] U.S. Department of Transportation, Federal Highway Administration, *Developing Traveler Information Systems Using the National ITS Architecture*, 1998.

[16] http://www.smarttrek.org/seattle_map.html.

[17] Institute of Transportation Engineers, *Intelligent Transportation Systems Primer*, Washington, D.C., 2001.

5
ITS Architecture

ITS architecture is a primary element for ITS planning. This chapter discusses the project and regional architecture and the importance of developing an ITS architecture. The major elements of an ITS architecture—concept of operations, user service requirements, logical architecture, and physical architecture—are also discussed. The chapter also describes a software tool that is available to aid in the development of an ITS or regional project architecture.

5.1 Regional and Project ITS Architecture

A regional ITS architecture can generally be defined as a framework for deploying ITS in a particular region. The region is typically comprised of several adjacent areas with multiple jurisdictions and stakeholders in common operations, where these stakeholders have agreed to share data. The architecture specifies how the different ITS components would interact with each other to help address regional transportation problems. An ITS regional architecture thus provides a general framework upon which to plan, design, deploy, and integrate systems in a particular region to realize selected objectives.

ITS architecture is different from design. A regional ITS architecture does not specify a design approach. Rather, it defines a general framework around which several design alternatives could be developed. These designs would still conform to a common architecture.

Project ITS architectures should be derived from the the regional ITS architecture, but in the absence of a regional ITS architecture, should be directly derived from the National ITS Architecture. If a regional ITS architecture exists, project ITS architectures are simply partitions or section of the regional ITS architecture. The primary purpose of the regional ITS architecture is to ensure that the projects ITS architectures interface seamlessly with each other.

5.2 Why Do We Need an ITS Architecture?

While an ITS architecture does not specify a design approach, it could provide many benefits to a region as it starts developing and deploying systems. ITS provides transportation professionals with a wide variety of options to address their needs. Given this, a common architecture and set of standards are needed to ensure that the systems deployed in the different regions around the country are "interoperable," or compatible with one another, and can communicate with each other.

Interoperability is very important in ITS. For example, a driver purchasing an in-vehicle display device in one state should be able to use his or her device to communicate with roadside equipment in another state. This could only be possible if the two systems in both states are compatible with each other at the system interface level. To facilitate this, an ITS architecture is needed to help identify the interfaces or information flows that need to be standardized. There are three types of interoperability, such as the following:

- International interoperability;
- National interoperability;
- Regional interoperability.

International interoperability is necessary for ITS applications that include sharing information across borders. Examples of systems that require international interoperability include commerical vehicle credentials necessary for crossing the border.

National interoperability implies standard system interfaces across the nation. Examples of systems that require national interoperability include commercial vehicle operations, emergency notifications, and electronic payment. Regional interoperability applies to systems that should interface within a region. Examples of systems that call for regional interoperability include incident management, emergency management, public transit management, and traveler information. Local interoperability is required for systems that are unique for a local scenario.

The real advantage of deploying ITS becomes more obvious when different ITS systems are capable of sharing and exchanging information with each other. For example, information collected by a traffic control system can be very useful to an emergency management center in trying to determine the fastest route for an emergency vehicle to reach an accident scene. At the same time, information collected by the emergency management center on incidents can be used by a traffic control center to adjust signal timings in response to the incident. The process of linking or connecting the different components of a system is typically referred to as *systems integration*. This will become a major issue as more and more ITS systems are deployed. A regional architecture is an important tool for the success of this process, since it describes how the different components of a system are envisioned to interact with one another.

Developing a regional ITS architecture helps define a logical implementation plan or blueprint as well as high-level functional requirements. The logical implementation plan helps the designer visualize a particular design in the overall system implementation plan as well as project phasing. The high-level functional requirements and interface requirements help define detailed functional requirements during the system design phase.

Resource sharing is an emerging trend in public-private partnerships. A resource-sharing initiative may call for a public organization to provide right-of-way or information in return for compensation in the form of infrastructure or services [1]. In ITS, many state departments of transportation are allowing the construction of cellular communications towers or the installation of communications cables in their rights-of-way. In return, the agencies are being compensated an agreed-upon sum of in-kind ITS program-related hardware and services.

The preferred approach to the resource-sharing process is first to define the region's telecommunications needs and develop potential communications architecture. This prepares the region to negotiate with private industry. By defining the architecture before entering into a resource-sharing agreement with private enterprises, stakeholders will have a better understanding of their needs before any negotiation occurs.

A regional ITS architecture development process starts with establishing consensus among stakeholders on sharing resources to achieve a common operation. While each agency in the region will interact with the deployed systems in a unique fashion to address its individual needs, sharing resources, especially communication systems, among different business units is likely to contribute to cost-effective interoperability. The ITS architecture will also help identify institutional interdependencies that exist in the region and how institutions can benefit from each other's activities.

5.3 Concept of Operations

A concept of operations is a prerequisite in developing an ITS architecture. The concept of operations identifies and describes who is going to perform what functions. A concept of operations for a regional ITS architecture is developed through a collaborative effort between the regional ITS stakeholders. It describes the roles and responsibilities of the stakeholders in the regional ITS operations by defining the functions of each stakeholder. The concept of operations is established through agreement(s) between the stakeholders.

There should be an open and all-inclusive process to engage a broad range of stakeholders to define the ITS framework for a region. This ensures that the regional ITS programs will be supported by all stakeholders and will receive the required interagency cooperation. Prior to defining the concept of operations, the need for regional or project ITS architecture should be assessed, the region or project should be defined, stakeholders, as well as a champion, should be identified, an inventory of the ITS-related systems should be made, and future ITS-related needs and serves should be defined. Potential stakeholders should include, but are not limited to, the following:

- State transportation department;
- *Metropolitan planning organizations* (MPOs);
- Local governmental agencies;
- Commercial vehicle operators;
- Land-side port-related agencies;
- Railroads authority;
- Transit agencies;
- Other state agencies;
- Information service providers;
- Airport authorities;
- Federal Highway Administration;
- Other federal agencies;
- Emergency management services;
- Toll administration;
- Regional planning authority.

Figure 5.1 provides an example of a concept of operations developed for the Northern Virginia area for personal travel services [2]. The personal travel services include pretrip traveler information, en route driver information, route guidance, electronic payment systems (not shown in the figure), emergency notification, and traveler services information. These services are described in Chapter 3.

ITS Architecture 97

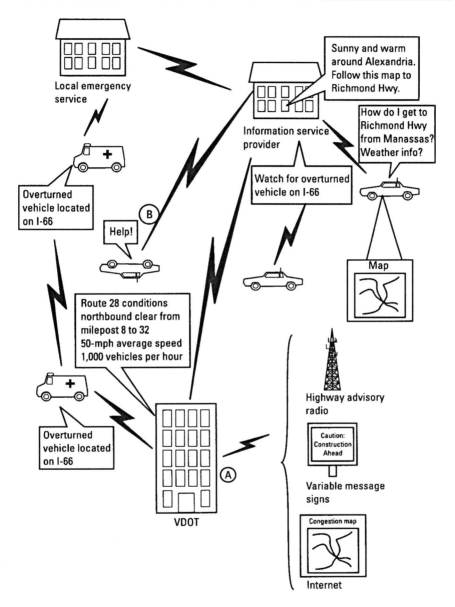

Figure 5.1 Example of a concept of operations for personal travel.[1] (*From:* [2]. © 1999 The Virginia Department of Transportation. Used with permission.)

---

1. Concept of operations: Stakeholders: Virginia Department of Transportation and Information Services Providers
   A. Virginia Department of Transportation (VDOT) will provide timely and accurate information on roadway conditions. Real-time information, such as traffic conditions, status of

5.4 National ITS Architecture

To provide guidance for regions deploying ITS, the U.S. Department of Transportation developed the national ITS architecture [3]. This architecture was developed over a period of 10 years by a broad cross section of the ITS community, including systems engineers, transportation practitioners, and technology specialists. The architecture represents a wealth of information that can be used to assist regional agencies in developing their own ITS architectures. The national ITS architecture documents, which are continuously being updated, are available on the Internet at http://www.iteris.com/itsarch. The national architecture, however, uses some terminology that is not commonly used by the transportation community. This section explains the basic terminology and concepts of the national architecture.

The national architecture contains the following five main components:

1. User services and user service requirements;
2. Logical architecture;
3. Physical architecture;
4. Equipment packages;
5. Market packages.

Each of these components is explained in the following sections.

5.4.1 User Services and User Service Requirements

The first step in developing an architecture for any type of a system is to define what the system is required to do. Consider, for example, the case of an architect developing an architecture for a house. The first task is obviously going to be to

---

park-and-ride lots, and construction activities, will be provided to motorists by VDOT through the use of highway advisory radio, variable message signs, highway helpline, and the Internet. Special transit information may be conveyed to motorists at strategic locations through variable message signs and highway advisory radio. In addition, VDOT will share data with information service providers, which will, in turn, provide packaged information through in-vehicle devices, dial-up services, PDAs, the Internet, television, and radio stations. B. Information service providers will support motorists equipped with an automated in-vehicle emergency reporting system. In-vehicle "Mayday" systems will allow quick and automated identification of disabled vehicles and serious accidents in remote areas. When triggered by a serious crash, or activated by the driver, the location of the event will be communicated to an information service provider, who will then forward the information to local emergency services and VDOT. The local emergency service will send appropriate help to the affected vehicle, and VDOT will support the effort by managing the traffic at the location of the incident.

identify the preferences and requirements of the house owner regarding the building's style, structure, number of rooms, and so on. The architect would then combine the information gained regarding the owner's preferences with existing information on the neighborhood environment and the pertinent building codes in an effort to develop an architecture for the house.

A similar process was followed in developing the national ITS architecture. The architecture development team started by receiving a set of 29 user services, defined by industry concensus and led by ITS America and U.S. DOT, that describe what the ITS system will do from the perspective of the user (i.e., travelers, drivers, transportation operators). To date, 32 user services have been identified and bundled into 8 user service areas, as discussed in Chapter 3.

A number of functions are needed to accomplish these user services. For example, to control traffic, the traffic would first need to be monitored and then the data information collected and used to optimize traffic flow. To reflect this, stakeholders led by ITS America divided each of the original 29 user services into a set of functional statements that describes the different functions required to accomplish each user service. These statements are called user service requirements, and each contains the word "shall." Table 5.1 shows examples of some of the user service requirements needed to accomplish the incident management user service.

To date, the National ITS Program Plan and its addendums list more than 1,000 user service requirements needed to implement ITS nationally. These statements constitute the fundamental requirements for the National ITS Architecture development effort, and hence the scope of the national ITS architecture itself.

If any of the functions that are to be performed by an agency cannot be mapped to the user service requirements, new user service requirements can be defined. In some instances, the user service requirements can be modified to make the requirements more reflective of a region's business practices.

Because a region may have unique and particular functions reflecting its business practices, there is sometimes no direct correlation to the user service requirements. In such cases, the user service requirements can be edited or supplemented with additional requirements. However, since the ITS standards program is based on the National ITS Architecture, which is driven by the user service requirements, caution should be exercised with regard to changing existing user service requirements.

An example of a modification to the National ITS Architecture is illustrated through a project conducted for the Virginia Department of Transportation (shown in Figure 5.2) [4]. Any text that has been added is noted in *italics*. The National ITS Architecture text that has been replaced shows the original text in ~~strikethrough~~ with the modifications in *italics* immediately next to the replaced text.

Table 5.1
Examples of User Service Requirements

1.7 SHALL include an incident management function. Incident management will identify incidents, formulate response actions, and support initiation and ongoing coordination of those response actions. Six major functions are provided, which are (1) schedule planned incidents, (2) identify incidents, (3) formulate response actions, (4) support coordinated implementation of response actions, (5) support initialization of response to actions, and (6) predict hazardous conditions.
1.7.1 Incident management SHALL provide an incident identification function to identify incidents.
1.7.1.1 The incident identification function SHALL include the capability to identify predicted incidents.
1.7.1.1.1 The incident identification function SHALL use information from the following types of sources, where available, to identify predicted incidents: • Traffic flow sensors; • Environmental sensors; • Public safety sources; • Media sources; • Weather information sources; • Transportation providers; • Sponsors of special events; • Hazardous condition prediction algorithms.
1.7.1.1.2 The incident identification function SHALL determine at least the following characteristics of each predicted incident: • Type; • Extent; • Severity; • Location; • Expected duration.

Source: [3].

Additional requirements are integrated with the original user service requirements. These are denoted in italics as well. The Virginia-specific requirements follow the numbering convention used in the original text, with the exception that they also include the letters "VA." For example, national architecture requirement "1.2.3.4" may be followed with a Virginia requirement "*1.2.3.5—VA.*" Note that the number of the Virginia-specific requirement is italicized as well.

ITS Architecture

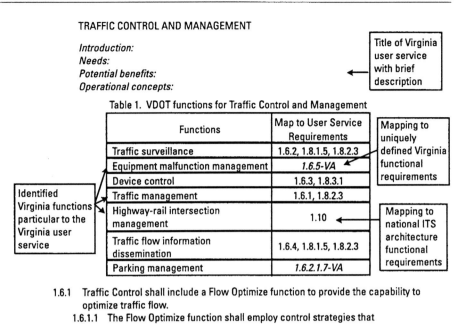

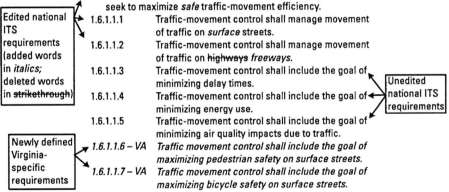

Figure 5.2 Editing the National ITS Program Plan user service requirements. (*From:* [4]. © 1999 Virginia Department of Transportation. Used with permission.)

5.4.2 Logical Architecture

The logical architecture provides a detailed description of the system's behavior that focuses on the functional processes involved and the information flows of a system. The description of a system's function typically includes a definition of the input data received by a function, the processing or manipulation of the data performed by the function, and the output data or information produced by the function. Consider, for example, the function needed to process traffic data. This function (shown in Figure 5.3) may be described as being responsible for

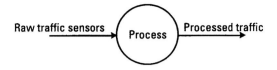

Figure 5.3 A high-level representation of a system's function.

receiving raw data from roadside traffic sensors processing this data (computing 30-second averages, for example), and then sending the processed data to other functions, such as those responsible for providing road traffic control.

In addition to describing the different functions of a system, the logical architecture describes the lower-end interaction of the different components comprising the regional ITS system. These components include the system's functions, the end users, and the existing or legacy systems. The interaction between system functions is exemplified by when traffic surveillance detectors identify congestion or incident and automatically call the nearest traffic control center. An example of the interaction between end users (personnel) and system functions occurs when a traffic engineer changes a signal controller-timing plan from a remote office. Finally, interaction between system functions and existing (legacy) systems occurs when the function responsible for special-event planning automatically calls the pavement management system (a legacy system) to retrieve a schedule of upcoming pavement improvement projects.

One important point to note about the logical architecture is the fact that it is not intended to identify how the functions will be implemented. Moreover, the logical architecture does not specify a particular technology. The implementation details are typically left for the later stages of the design phase.

As previously described, the logical architecture is intended to describe the different functions or processes of a system and the data inputs and outputs of each function. In the national ITS architecture, the processes and data flows are grouped and represented graphically in the form of *data flow diagrams* (DFDs), which are organized in several levels of detail. In these diagrams, processes are represented as circles, and data flows are indicated by arrows, based on Hatley and Pirbhai's structured design methodology [5].

Figure 5.4 shows the processes of the National ITS Architecture at the highest level. Each process is then broken down into more basic subprocesses for each level.

Figure 5.5 illustrates this concept. The Manage Traffic process is first broken into six more fundamental subprocesses, one of which is the Provide Traffic Surveillance process. The Provide Traffic Surveillance subprocess is then broken

ITS Architecture 103

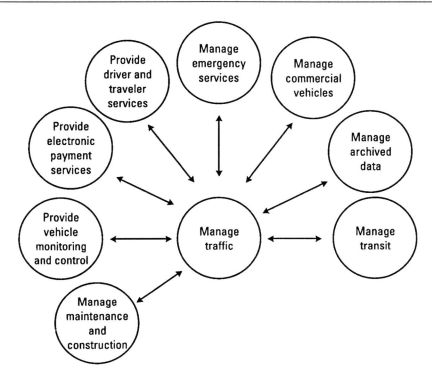

Figure 5.4 The highest level of the national ITS logical architecture. (*After:* [3].)

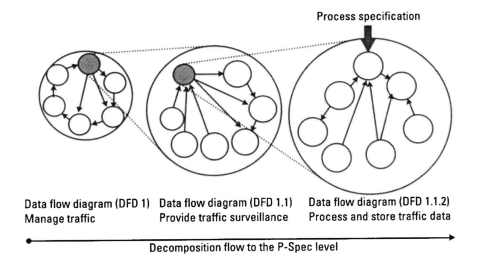

Figure 5.5 Process decomposition into elemental functions. (*After:* [3].)

down into seven subprocesses, and so on. The process specifications (or P-specs) contain the lowest level of detail. The P-specs, such as the Process Traffic Data P-Spec, as pointed by arrow, can thus be defined as the most elemental functions (those that cannot be further broken down) that need to be performed to satisfy user service requirements.[2]

5.4.3 Physical Architecture

The physical architecture builds on the logical architecture foundation by defining the physical subsystems and architecture flows. System functions from the logical architecture that serve the same need are grouped into subsystems. For instance, a traffic management subsystem may contain the traffic surveillance, fault detection, and device control functions. The subsystems categorize the functions so that a physical entity can be developed that delivers the functions. In addition, the data flows from the logical architecture are grouped together into physical architecture flows. Figure 5.6 illustrates the procedure for developing the physical architecture based on the logical architecture definition. As can be seen, functions A and B were assigned to subsystem A, whereas functions C and D were assigned to subsystem B. The data flows of the logical architecture were also combined to define the interface between the two subsystems.

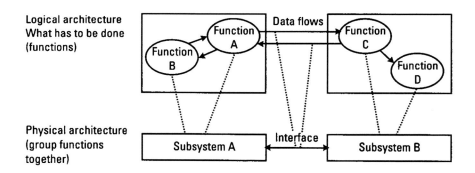

Figure 5.6 Moving from the logical to the physical architecture. (*After:* [3].)

---

2. Definition of the Process Data P-Spec [3]: "Overview: This process shall receive and process data from sensors (both traffic and environmental) at the roadway. The process distributes data to Provide Device Control processes that are responsible for freeway, highway rail intersections, parking lot and surface street management. It also sends the data to another Provide Traffic Surveillance process for loading into the stores of current and long-term data. This process distributes environmental sensor data to other processes in Manage Traffic as well as the process that is responsible for monitoring vehicle speed. Information about various sensors to aid in this processing and distribution of data is accessed from the data store static_data_for_sensor_processing.

In the physical architecture, interfaces exist between different subsystems, between subsystems and terminators. The physical architecture and logical architecture comprise the functional definition, or core, of the National ITS Architecture.

In the physical architecture, the logical architecture's processes are allocated to the system's physical components (subsystems), and the data flows are aggregated into physical architecture flows. Figure 5.7 shows the 21 subsystems that comprise the national ITS physical architecture. The 21 subsystems can be grouped into up to four basic classes: the centers; the roadside; the vehicles; and the travelers. The subsystems represent aggregations of functions that serve the same transportation need and closely correspond to the physical elements of transportation management systems. For example, the traffic management subsystem (one of the 10 centers' subsystems) represents the functions typically performed by a traffic control center. The roadway subsystem (one of the four roadside subsystems) is comprised of roadside devices such as traffic controllers, traffic signals, loop detectors, and CCTV cameras. The vehicles subsystem corresponds to the five different types of vehicles using the transportation system—passenger cars, transit vehicles, commercial vehicles (trucks), and emergency vehicles (ambulances, police cars, and fire trucks) and maintenance and construction vehicles. The travelers subsystem represents the different ways a traveler can access information on the status of the transportation system.

Figure 5.7 also shows the different communications classifications (indicated in gray) connecting the different subsystems. As can be seen, four different types of communications systems are used: wireline communications; wide-area wireless communications; *dedicated short-range communications* (DSRC); and vehicle-to-vehicle communications. Wireline communications can be used to connect the center's subsystem to the roadside subsystem; an example includes the fiber-optics networks used to connect traffic control centers to the freeway loops and VMSs. Wide-area wireless communications can be used to connect remote travelers to the different components of the transportation system. DSRC involves communications between vehicles and roadside equipment; an example includes communications between a vehicle and a roadside reader. Finally, vehicle-to-vehicle communications refer to communications between the vehicles—a feature of the automated highway concept.

The interfaces of an ITS system at the physical architecture level should map to national standards such as the *National Transportation Communications for ITS Protocol* (NTCIP), which will be discussed in more detail in Chapter 7. These standards arose from the need to ensure a system's compatibility and interoperability, so that system developers do not have to rely on proprietary technology from a manufacturer. The standards' conformity to interfaces is an important issue because new standards are currently being identified as the ITS industry grows.

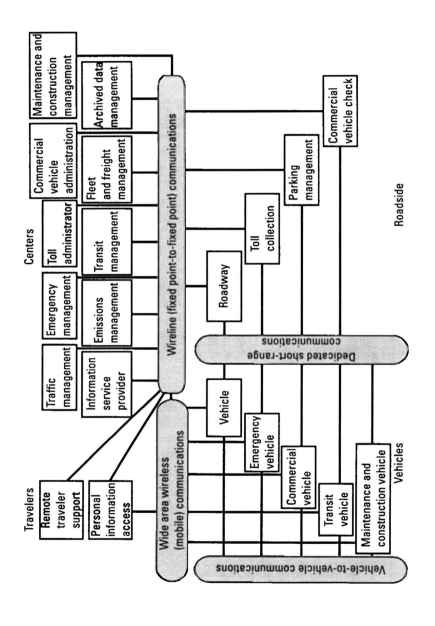

Figure 5.7 National ITS Architecture subsystems and communications. (*After:* [3].)

5.4.4 Equipment Packages

While all information needed to define the national ITS system is contained within the logical and physical architectures previously described, the architecture development team introduced the concepts of equipment and market packages in an effort to provide a more deployment-oriented perspective to the National ITS Architecture. An equipment package refers to a collection of similar functions (P-specs) of a particular subsystem that can be grouped together and implemented by a typical package of hardware and software capabilities. In total, the national ITS architecture defines 176 equipment packages. An example of an equipment package is the TMC signal control equipment package, which provides the capability for traffic managers to monitor and manage traffic flow at signalized intersections. This equipment package, which belongs to the traffic management subsystem, is comprised of five P-specs: Provide Traffic Operations Personnel Traffic Interface; Process Traffic Data; Select Strategy; Determine Indicator State for Road Management; and Output Control Data for Roads. The TMC signal control equipment package is in the Surface Street Control and Emergency Routing market packages.

5.4.5 Market Packages

Market packages define a set of deployment-oriented ITS service blocks. Each market package is defined by a set of one or more equipment packages that are required to work together to provide a particular transportation service or solve a particular transportation problem. These equipment packages typically belong to more than one subsystem; consequently, the market package definition also shows the major architecture flows occurring between these subsystems. Version 4.0 of the national architecture includes a total of 75 market packages categorized into the following eight clusters:

1. Traffic management;
2. Public transportation;
3. Traveler information;
4. Vehicle safety;
5. Commercial vehicle operations;
6. Emergency management;
7. Archived data management;
8. Maintenance and construction management.

Figure 5.8 shows an example of a market package. The figure shows that the surface street control market package is made up of four equipment packages

108   Fundamentals of Intelligent Transportation Systems Planning

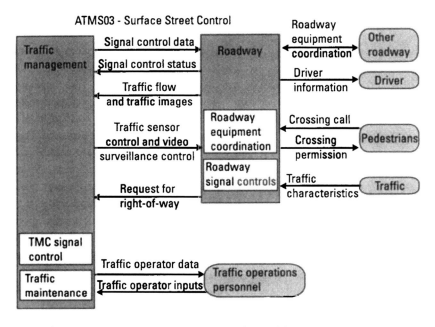

Figure 5.8 Surface street control market package. (*After:* [3].)

(indicated by the white rectangles). Two equipment packages reside in the traffic management subsystem and two reside in the roadway subsystem. The figure also shows the major architecture flows between the two subsystems and between the subsystems and terminators (indicated by the arrows).

The use of the market packages provides a convenient means to develop a regional ITS architecture by first identifying the set of market packages that satisfy the user requirements. The equipment packages within each subsystem could then be combined from each of the selected market packages. Finally, the architecture flows between the same subsystems are combined into the regional ITS architecture from each of the market packages.

5.5 Proposed Procedure for Developing a Regional ITS Architecture

Developing a regional ITS architecture is typically conducted as a part of the process to develop an ITS plan for a particular region, which will be discussed in Chapter 6. An excellent reference containing a typical process for creating a regional ITS architecture that references the Final Rule/Policy is the "Regional ITS Guidance—Developing, Using and Maintaining and ITS Architecture for Your Region" document [6].

5.6 Architecture Development Tool

To facilitate the application of the National ITS Architecture, the Federal Highway Administration has developed an interactive software package called Turbo Architecture that generates regional and/or project ITS architecture based on the user's input to a specific questionnaire [7]. Turbo Architecture utilizes user inputs and the National ITS Architecture to provide a physical representation of regional or project ITS architecture in tabular and graphical outputs. Users of Turbo Architecture, however, should be familiar with the National ITS Architecture to apply this tool effectively. The introductory questionnaire only helps the user begin the development of ITS architecture; substantial additional input and customization is required.

As mentioned, this tool allows the user to develop a regional or project ITS architecture by receiving input from users on a series of questionnaires (referred to as "interview" in the software). The questionnaire gathers information on existing or planned ITS infrastructure and services related to the selected programs, which includes the following:

- Electronic clearance of commercial vehicles;
- Emergency management;
- Electronic tolling;
- Public transportation;
- Regional traveler information;
- Freeway management;
- Arterial or traffic management.

The tool allows the user to develop tailored reports and generate architecture diagrams, including a sausage diagram (similar to Figure 5.7), an interconnect diagram (showing the different ITS subsystems, terminators and their connections), and an architecture flow diagram (showing the information that needs to be exchanged among the different subsystems). Figure 5.9 shows an interconnect diagram developed using the software with emergency management, freeway management, and arterial or traffic management chosen as the major services.

5.7 Conclusions

In this chapter, the significance and benefits of developing a regional ITS architecture to guide the deployment of ITS technologies were discussed. The

110 Fundamentals of Intelligent Transportation Systems Planning

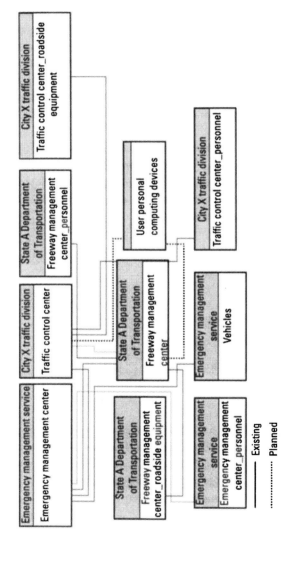

Figure 5.9 Interconnect diagram generated by the Turbo Architecture.

different components of the National ITS Architecture were then described, and a proposed process for developing a regional ITS architecture was outlined. Software to aid in the application of the National ITS Architecture was also described. Chapter 6 will discuss how the ITS architecture can be a tool to ITS planning. Chapter 7 will discuss how the National ITS Architecture supports the development of ITS standards.

Review Questions

1. Discuss some of the benefits behind developing a regional ITS architecture.
2. What are the basic elements or components of a regional ITS architecture?
3. Distinguish between the logical and the physical architecture.
4. The logical architecture describes the interaction of the different components making up the regional ITS system, including the system's functions, end users, and existing or legacy systems. Give some examples for such interactions.
5. In the context of the National ITS Architecture, explain what is meant by *equipment package* and *market package*.
6. Access the National ITS Architecture available on the Internet at http://www.iteris.com/itsarch. List 10 different ITS market packages and give a brief description for each.
7. Develop a physical architecture for a system that combines the following three market packages: network surveillance, traffic information dissemination, and incident management. Refer to the National ITS Architecture at http://www.iteris.com/itsarch for details regarding these three market packages.

References

[1] Pol, J., M. Chowdhury, and S. Malek, "Resource Sharing Implications for an ITS Physical Architecture," *69th ITE Annual Meeting*, Las Vegas, NV, 1999.

[2] Virginia Department of Transportation (VDOT), *Smart Travel Program in the Virginia Department of Transportation, Northern Virginia District Strategic Plan*, July 1999.

[3] Federal Highway Administration, *National ITS Architecture*, updated 2002.

[4] Virginia Department of Transportation (VDOT), *Smart Travel User Service Definitions Reference*, July 15, 1999.

[5] Hatley, D. J., and I. A. Pirbhai, *Strategies for Real-Time System Specification*, New York: Dorset House Publishing, 1988.

[6] *National ITS Architecture Team, Regional ITS Architecture Guidance—Developing, Using and Maintaining an ITS Architecture for Your Region*, Report No. FHWA-0P-02-024, October 12, 2001.

[7] Federal Highway Administration, *Turbo Architecture*, Version 2.0, 2002.

6

ITS Planning

This chapter discusses opportunities to integrate ITS into the transportation planning process. Fundamentals of traditional transportation and ITS planning are introduced to show where these processes could be integrated. In addition, this chapter presents an overview of how the National ITS Architecture could be a primary resource for ITS planning. Two methods are presented for ITS planning, both of which require input from the National ITS Architecture.

6.1 Transportation Planning and ITS

Transportation planning helps shape a well-balanced transportation system that can meet future demands. Future traffic demands and required supply in terms of roadways and transportation modes are estimated through transportation planning. This exercise helps develop needed projects to satisfy future demands.

Traditionally, a transportation plan identifies projects for different modes of transportation, such as roadways and transit, for a 20-year period. A short-term document called the *Transportation Improvement Program* (TIP) is developed based on the 20-year transportation plan. The TIP, which covers a 3-year period and is updated every 2 years, lists a group of priority projects to be conducted each year.

Transportation planning is an iterative stepwise process. Figure 6.1 shows the typical major steps in the urban transportation planning process. The process starts with defining goals and objectives, which should be developed based on public opinion and a regional vision for the transportation system. The next step is performing an inventory of existing conditions, such as land use, transportation facilities and their usage, travel patterns, and resources. Data on

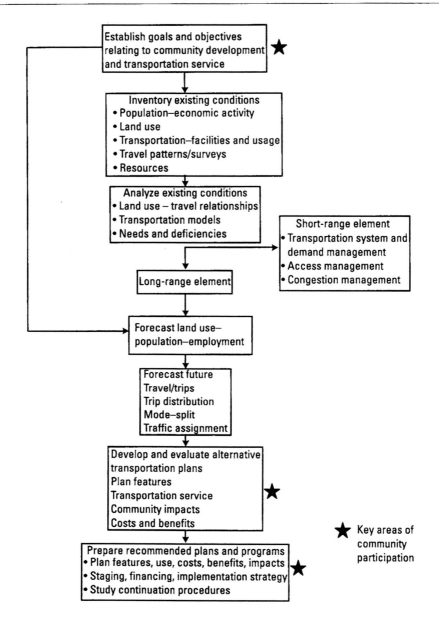

Figure 6.1 Urban transportation planning process. (*From:* [1]. © 1999 Institute of Transportation Engineers. Used with permission.)

existing conditions is analyzed in the next step using analytical relationships or transportation models to identify deficiencies and needs. Based on the needs, short- and long-range solution strategies or projects are identified.

The short-range elements include operational improvement strategies that could be implemented in the immediate future, such as demand, access, and congestion management. The long-range elements ultimately result in improvement projects and programs that require significant resources and time to implement. The identification of long-range programs and plans goes through another process where future land use and traffic demand between different geographical locations within a planning area are identified. Based on the forecasted travel demand, alternative projects are identified to meet the demand. These alternatives are evaluated with participation from the community. Ultimately, particular alternatives are selected, also with participation from the community, based on costs, benefits, and any other impacts. Once long-range projects are selected, implementation strategies are developed, along with a detailed work plan and schedule and the identification of funding sources.

The ITS planning process differs somewhat from the planning process for a traditional transportation infrastructure. ITS or advanced systems approach a transportation problem in a different manner. Table 6.1 shows a comparison between the conventional approach and the ITS approach to solving various transportation problems. Computers, communications, and software, which are the primary components of an ITS project, are complex and their underlying technology is rapidly advancing. They are also notoriously difficult to change or modify once in place. Because it is so difficult to modify technology systems, ITS planning must be sufficiently detailed to ensure that the ITS installed today has the capability designed into the original systems for future expansion. ITS planning should envision future transportation service needs, including geographic and functional needs, and envision complete ITS functions to meet those needs.

ITS has the unique capability to integrate different modes of transportation, such as public auto, transit, and infrastructure elements through communications and control. This multimodal integration potential provides a great opportunity for planning across modes.

The statewide transportation plan, which covers statewide transportation programs and projects, is coordinated with the urban transportation plan. ITS should be integrated with urban as well as statewide transportation plans. Specific Federal requirements are imposed on statewide and metropolitan transportation planning. Table 6.2 shows the requirements for a statewide transportation plan and relevant elements that should be considered for including ITS in the plan. Similarly, Table 6.3 shows the requirements for a metropolitan or urban transportation plan and relevant elements that should be considered for including ITS in the plan.

One of the key elements in the 1998 Transportation Efficiency Act for the Twenty-first Century (TEA-21) was a recommendation to mainstream ITS into the regular federal-aid program. TEA-21 emphasized operations and

Table 6.1
Relationships Between Problems, Conventional Approaches, and Advanced Technology Approaches

Problem	Possible Solutions	Conventional Approach	Advanced Systems Approach
Traffic congestion	Increase roadway capacity (vehicular throughput)	New roads New lanes	Advanced traffic control Incident management Electronic toll collection Advanced vehicle systems (reduce headway)
	Increase passenger throughput	HOV lanes Car pooling Fixed route transit	Real-time ride matching Integrate transit and feeder services Flexible mode transit Personalize public transit
	Reduce demand	Flex-time programs	Telecommuting Other telesubstitutions Transportation pricing
Lack of mobility and accessibility	Provide user-friendly access to quality transportation services	Expand fixed-route transit and para-transit service Radio and TV traffic reports	Multimodal pretrip and en route traveler information services Respond dynamically to changing demand Personalize public transportation Common, enhanced fare card
Disconnected transportation modes	Improve intermodality	Interagency agreements	Regional transportation management systems Regional transportation information clearinghouse Disseminate multimode information pretrip and en route
Severe budgetary constraints	Use existing funding efficiently	Existing funding authorization and selection processes	Privatize market packages Public-private partnerships Barter right-of-way Advanced maintenance strategies
	Leverage new funding sources		Increased emphasis on fee-for-use services
Transportation following emergencies	Improve disaster response plans	Review and improve existing emergency plans	Establish emergency response center (ERC) Internetwork ERC with law enforcement emergency units, traffic management, transit

Table 6.1 (continued)

Problem	Possible Solutions	Conventional Approach	Advanced Systems Approach
Traffic accidents, injuries, and fatalities	Improve safety	Improve roadway geometry	Partially and fully automated vehicle control systems
		Improve sight distance	Intersection collision avoidance
			Automated warning systems
		Traffic signals, protected left-hand turns at intersections	Vehicle condition monitoring
			Driver condition monitoring
			Driver vision enhancement
		Grade separate crossings	Advanced grade crossing systems
		Driver training	Automated detection of adverse weather and road conditions, vehicle warning, and road crew notification
		Sobriety check points	
		Lighten dark roads to improve visibility/ better lighting	
			Automated emergency notification
		Reduce speed limits/ post warnings in problem areas	

Source: [2].

management planning in the transportation planning process as a strategy to mainstream ITS. Usually, MPOs serve as the regional planner of the transportation system. These agencies conduct the fundamental steps of the transportation planning process and develop future projects. MPOs can be an effective entity to integrate ITS into the overall transportation planning process. Various regional MPOs have already been active in the process of recognizing or integrating ITS in the overall transportation planning framework. For example, the Metropolitan Transportation Commission of San Francisco, the local MPO, has developed a regional transportation management plan where ITS was a major tool. Similarly, the Metropolitan Washington Council of Governments, the MPO for the Washington, D.C., area, has adopted ITS as one of the key tools for solving future mobility problems.

6.2 Planning and the National ITS Architecture

The National ITS Architecture is a useful tool for integrating ITS into the traditional planning process. The National ITS Architecture defines a

Table 6.2
Requirements for Statewide Transportation Plan and ITS

Requirement	Considerations for ITS
Be intermodal and statewide in scope in order to facilitate the efficient movement of people and goods.	ITS could be considered an important component for maintaining system efficiency. ITS cuts across multiple modes.
Be reasonably consistent in time horizon among its elements, but cover a period of at least 20 years.	Encourages examination of long-term approach to ITS. However, this is difficult given rapid changes in technology.
Contain, as an element, a plan for bicycle transportation, pedestrian walkways, and trails that is appropriately interconnected with other modes.	Not substantially relevant to ITS.
Be coordinated with metropolitan transportation plans.	Statewide ITS elements should be coordinated with metropolitan ITS elements.
Reference, summarize, or contain any applicable short-range planning studies, strategic planning and/or policy studies, transportation need studies, management system reports, and any statement of policies, goals, and objectives regarding issues such as transportation, economic development, housing, social and environmental effects, energy, and so on, that were significant to the development of the plan.	ITS strategic assessment, corridor, or subarea studies that involve ITS and ITS policy statements should be referenced or incorporated into the transportation plan. ITS-related plans should be packaged in such a way that certain summary elements could be easily incorporated into the transportation plan.
Reference, summarize, or contain information on the availability of financial and other resources needed to carry out the plan.	Any financial resources that are generated through ITS applications (e.g., from private sector, user fees, and so on) should be included.

Source: [3].

comprehensive set of data that should be shared between various agencies that are stakeholders in the regional transportation network. With the knowledge of what data must be exchanged and shared between various agencies, these agencies develop a common interest in cooperative planning efforts between all transportation projects with and without ITS components. Figure 6.2 shows the application of the National ITS Architecture in the planning phase in the project life cycle.

The planning process could benefit the most, among other major activities in the life cycle, from National ITS Architecture resources. Various elements of the National ITS Architecture could be used in the planning process. Section 6.3 shows how the user service requirements and market packages could be used in the planning process.

Table 6.3
Requirements for Urban Transportation Plan and ITS

Requirement	Considerations for ITS
Include both long-range and short-range strategies or actions that lead to the development of an integrated intermodal transportation system that facilitates the efficient movement of people and goods.	Even though many ITS actions may be short- to mid-term, they should be included in the transportation plan. The plan covers all time periods.
Identify the projected transportation demands of persons and goods in the metropolitan planning area over the period of the plan.	Estimates of benefit for ITS should take into account the future, not just the present.
Identify adopted congestion management strategies.	ITS is one of a number of congestion management strategies, oriented toward the more efficient use of existing facilities.
Identify pedestrian walkway and bicycle transportation facilities.	Not substantially relevant to ITS.
Reflect the consideration given to the results of the management systems.	Any ITS strategy evaluation and project recommendations conducted through the management systems should be reflected.
Assess capital investment and other measures necessary to preserve the existing transportation system and make the most efficient use of existing transportation facilities to relieve vehicular congestion and enhance the mobility of people and goods.	ITS could represent one form of capital investment to preserve and make the most efficient use of the existing transportation system.
Include design concept and scope descriptions of all exiting and proposed transportation facilities in sufficient detail in nonattainment and maintenance areas to permit conformity determinations.	If ITS is to be used as an element of demonstrating transportation conformity, it must be specified in sufficient detail to be evaluated under those requirements.
Reflect a multimodal evaluation of the transportation, socioeconomic, environmental, and financial impact of the overall plan, including all major transportation investments.	To the extent possible, the ITS elements should be included in the transportation, environmental, and financial evaluation. ITS is not generally relevant to socioeconomic factors.
Include a financial plan that demonstrates the consistency of proposed transportation investments with already available and projected sources of revenue.	Cost of ITS elements will need to be computed and balanced against total revenue. ITS will need to include operations and maintenance elements.

Source: [3].

6.3 Planning for ITS

ITS is a function of information and communication technologies, which are changing at a rapid pace. As the world moves through the peak of the

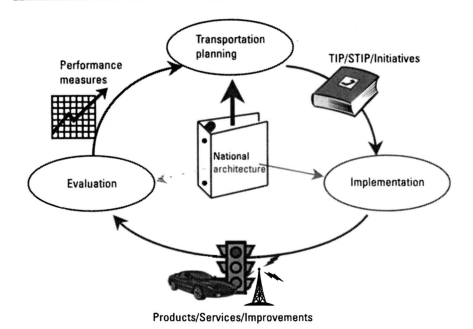

Figure 6.2 Transportation planning in the project life cycle. (*After:* [2].)

information superhighway, newer technologies will enable a region to obtain better functionality from the deployed systems. At the same time that a region moves toward the future, it will face different and newer challenges in transporting goods and people. Thus, ITS planning should be strategic in nature. It should be a "living" document that should be revisited periodically to reflect the recent needs and solution strategies and tools. This plan should be revised as a region gains significant experience in ITS through existing and future projects.

A study sponsored by the FHWA recommended mainstreaming ITS into planning, program, and project development processes, such as the Transportation Plan, TIP, and the *Congestion Management System* (CMS), to attract and sustain support for ITS [4]. The study suggested the following three approaches for institutionalizing ITS into the activities of a planning organization:

1. Establish ongoing research and analysis activities related to ITS;
2. Establish an ITS task force or committee;
3. Develop an ITS strategic plan.

These steps, according to the study, should help include ITS in routine transportation-related decision making.

The following sections presents two alternatives that could be considered for ITS planning. The first is based on the market package approach (see Chapter 5) and the second is based on traceability of projects to vision, goals, and objectives.

6.3.1 Market Package-Based ITS Planning Process

Developing a regional ITS architecture is typically conducted as a part of the process to develop a strategic ITS deployment plan for a particular region. Figure 6.3 shows the ITS planning process and the strategic deployment plan, proposed by the FHWA, that can be used to develop the regional ITS architecture [2]. Market packages are discussed in Chapter 5.

The first step is to identify the main ITS stakeholders in the region. This typically includes members from the department of transportation, the MPO, the department of public works, the police department, and the trucking industry, among others. With the stakeholders identified, the next step is to develop a set of vision, goals, and objectives statements. The definition of these statements allows agencies to define the ITS areas they would like to emphasize as they relate to the broader community needs. The study should also attempt to identify the main transportation problems and needs.

Following the definition of the vision and goals and the identification of the region's transportation problems, the process proceeds to screen the 75 *market* packages defined by the National ITS Architecture, in an effort to identify those packages most applicable to the region. There are many reasons why a particular market package may not be applicable to a given region. For example, a package may not relate to the region's identified problems or may not address its goals and objectives. A package may also be excluded on financial grounds, if its expected costs are likely to fall beyond the financial capabilities of the region.

With the set of market packages identified, the regional architecture can be developed as described. The equipment packages within each subsystem are combined from each of the selected market packages, and the data flows between the same subsystems are combined from each of the market packages. The functional requirements of the regional ITS system can be easily identified from the P-specs that are included in each of the equipment packages.

6.3.2 Traceability-Based ITS Planning Process

The ITS planning process follows a systems engineering approach to develop a deployment plan, which is a top-down approach that describes the vision, goals, objectives, and functions that support holistic systems development. The vision portrays the services ITS can provide to enhance the efficiency of the transportation network. Goals that provide initial direction on how to proceed

122 Fundamentals of Intelligent Transportation Systems Planning

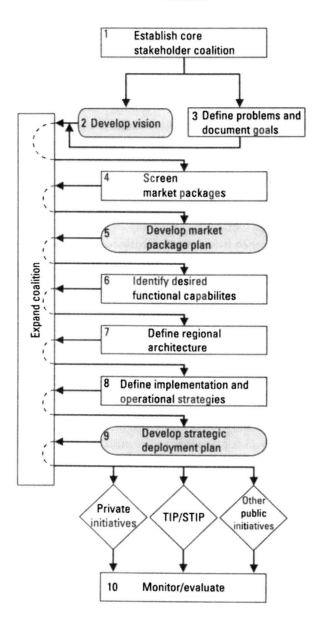

Figure 6.3 Market package-based ITS planning process. (*After:* [2].)

toward the vision are then derived. Many goals are directly connected to on-street safety and operations needs. Objectives further clarify goals and provide a more specific direction that is used to define functions. Selected functions are aggregated to define projects. This logical hierarchy, as shown in Figure 6.4, allows any project to be traced back to the goals and objectives, ensuring consistency with the state ITS program, as well as other long-range planning documents.

The "functions" of the ITS program are equivalent to the user service requirements of the National ITS Architecture (see Chapter 3). Thus, the ITS planning process that defines the required functionality of the regional system relates directly to the user services of the National ITS Architecture. This relationship ensures consistency with the national architecture, as required for federal aid eligibility.

Each function is again mapped to the National ITS Architecture user service requirements or regional ITS user service requirements, which, if they exist, have been adapted from the national requirements. These mapping exercises will verify whether the functions identified can be traced to national or regional ITS user service requirements.

To develop the long-range program, functions derived from the strategic planning process should be compared to functions provided by current systems in the region. Identifying gaps between current deployment and the functions envisioned by a strategic plan provides the project list, which can then be incorporated in the TIP.

In a deployment plan, each project at specific locations is identified based on potential functionality needs as identified from the traceability-based approach described earlier. Following this, appropriate technologies that suit the needs of candidate locations are identified. For each project, the project ITS architecture is identified in addition to the cost, purpose or benefits, and staffing requirements. The project ITS architectures show the physical entities of the project and the data that will be shared between these entities.

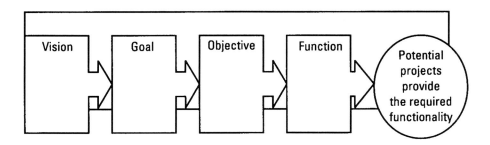

Figure 6.4 Traceability-based ITS planning framework. (*After:* [5].)

6.4 Case Study: Northern Virginia ITS Planning Exercise Using the Traceability-Based Approach

The following case study follows the development of an ITS strategic plan for the *Virginia Department of Transportation's Northern Virginia District* (VDOT NOVA) [5]. ITS is referred to as "Smart Travel" by VDOT. This strategic plan was developed after identifying the Smart Travel vision, goals, objectives, and functions for the VDOT NOVA. The vision portrays the services Smart Travel could offer to the district's transportation network. The vision was initially identified based on several regional studies, such as the Statewide Smart Travel Business Plan, the Northern Virginia Early Deployment Plan, and VDOT's Strategic Plan for the Twenty-first Century. Goals, which were derived from the vision, provide initial direction on how to proceed to the vision. Objectives further clarify goals, provide a more specific direction, and were used to define the functions that would ultimately lead to the vision. Selected functions were aggregated to define a project. The hierarchy of action planning allowed any project to be traced to the vision, thus enabling the vision to be realized through project deployment.

The strategic plan allocates functions to a region's existing institutional framework that identifies "who does what and where." The understanding of "who does what and where" will help develop the logical and physical architecture for a region. The logical architecture provides detail on the behavior of the system that is to be deployed. The physical architecture details the data that must be exchanged between physical entities, the means of exchanging data among them, and interface requirements between the physical entities. A regional ITS architecture will demonstrate interfaces and integration requirements between various stakeholder agencies within a region.

The planning effort by VDOT also developed a summary of Smart Travel activities in the region, which included a list of existing and planned ITS systems in the Northern Virginia region. A workshop was conducted with the leadership and staff of the VDOT Northern Virginia district to receive input on the inventory of existing systems and discuss responsibilities for deploying, operating, and maintaining them.

After identifying the differences between the functions offered by existing systems and the functions envisioned in the strategic plan, new projects were identified. Detailed programmatic information such as functional requirements and cost were developed for critical projects (which could be implemented in the short term).

The strategic plan was meant to be the foundation for the VDOT NOVA ITS program plan, supporting the development of potential projects that would forward the district's ITS program. This document helped the state focus its ITS program on a common vision and then identify how to reach that vision.

The ITS vision is the first element in a hierarchy of action planning for VDOT NOVA. The vision guides the program as goals are mapped to visions. Objectives to reach the goals are subsequently identified. The objectives are used to define functions to be undertaken by VDOT NOVA, and selected functions are aggregated to define a project. This hierarchy of action planning enables a traceability of any project to the vision, which ensures the realization of the ITS vision through project deployment.

Figures 6.5 through 6.10 show examples of how projects were mapped to functions, functions to objectives, objectives to goals, and goals to vision.

6.5 Integrating ITS into Transportation Planning

Integrating ITS into the transportation planning process requires overcoming some obstacles and some changes in the business practices of many institutions. A report published by the U.S. DOT identified the following as the major challenges in mainstreaming ITS into the everyday operations of transportation agencies (U.S. DOT) [3]:

- *Institutional coordination and cooperation.* The main benefits of ITS are sharing information and data. Stakeholders must cooperate and coordinate to ensure optimal success of any ITS project.

- *Technical compatibility between and among ITS projects.* Various projects must be able to share information. An ITS architecture would ensure interoperability between and among ITS projects.

- *Human resource needs and training.* Deployment, operation, and maintenance of ITS require special skills. Transportation agencies must make an investment in building the technical capacity of its employees to meet these demands.

- *Financial constraints and opportunities to involve the private sector.* ITS could not be implemented in the mainstream without private-sector resources. Public agencies must find ways to engage and attract the private sector to ITS.

Most public agencies are aware of the challenges in mainstreaming ITS. With the enormous promise ITS holds for the transportation system, public agencies have to be constantly engaged in meeting these challenges. In the past, ITS funding was provided separately; consequently, ITS activities were carried out independent of other transportation projects. Mainstreaming ITS into the transportation planning process where ITS projects are part of traditional

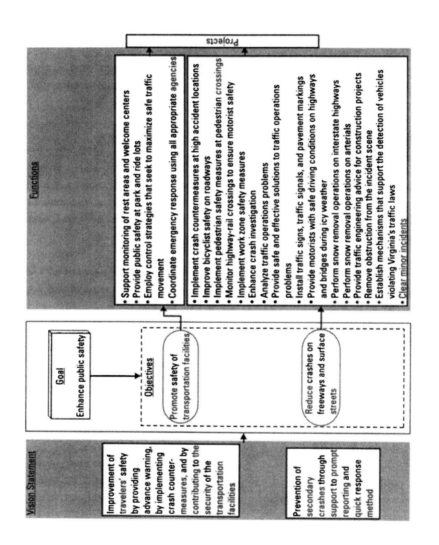

Figure 6.5 Traceability for the goal "enhance public safety." (*From:* [5]. © 2000 Virginia Department of Transportation. Used with permission.)

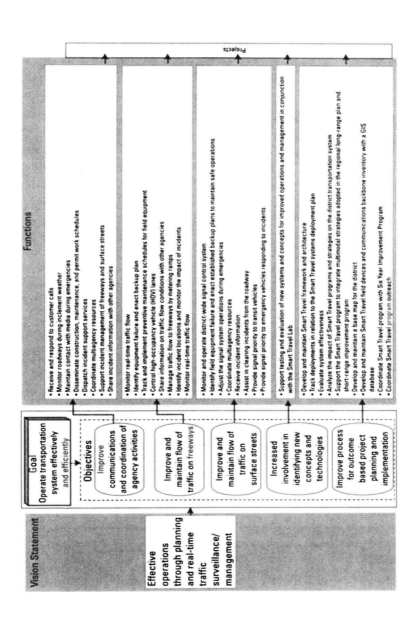

Figure 6.6 Traceability for the goal "operate the transportation system more effectively and efficiently." (From: [5]. © 2000 Virginia Department of Transportation. Used with permission.)

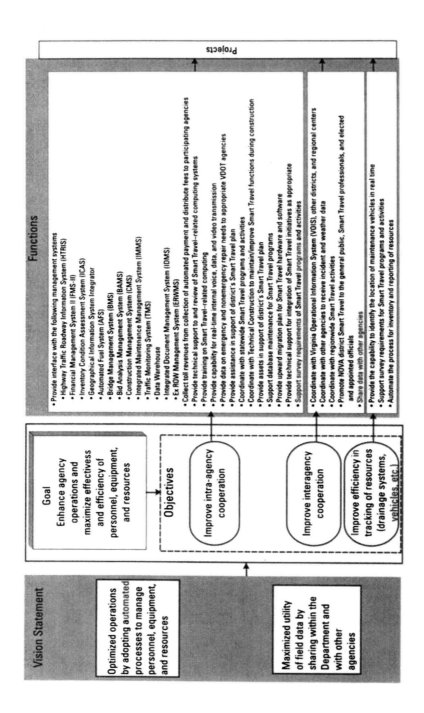

Figure 6.7 Traceability for the goal "enhance agency operations and maximize effectiveness and efficiency of personnel, equipment, and resources." (*From:* [5]. © 2000 Virginia Department of Transportation. Used with permission.)

ITS Planning

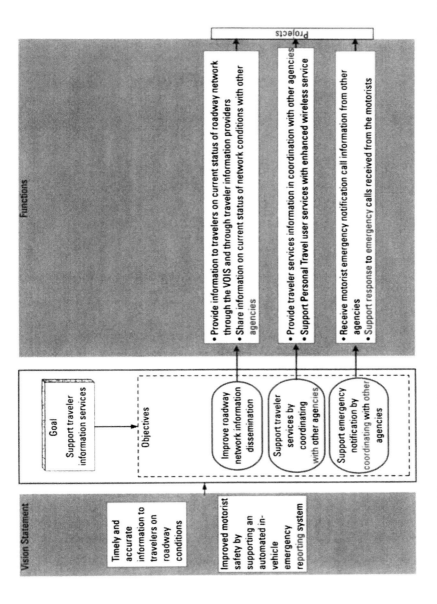

Figure 6.8 Traceability for the goal "support traveler information services." (*From:* [5]. © 2000 Virginia Department of Transportation. Used with permission.)

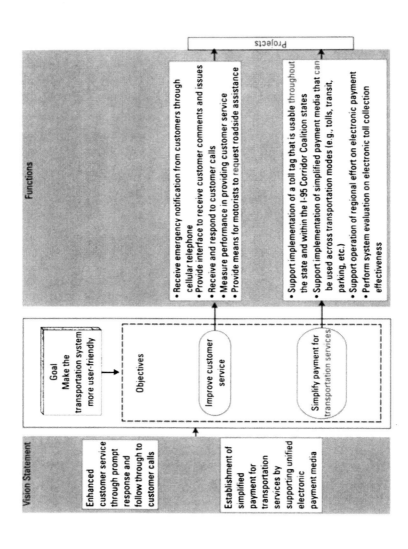

Figure 6.9 Traceability for the goal "make the transportation system more user-friendly." (*From:* [5]. © 2000 Virginia Department of Transportation. Used with permission.)

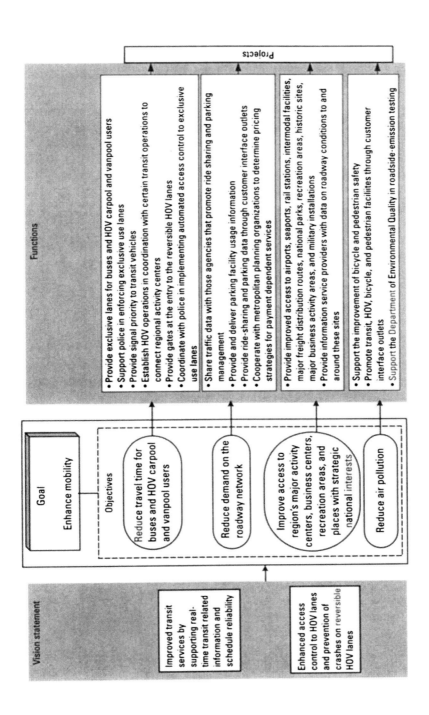

Figure 6.10 Traceability for the goal "enhance mobility." (*From:* [5]. © 2000 Virginia Department of Transportation. Used with permission.)

transportation programs on the local and state level is key to achieving the best output from transportation investments.

The *Intelligent Transportation System Deployment Analysis System* (IDAS), developed under the sponsorship of the U.S. *Department of Transportation* (DOT), can be a useful tool for integrating ITS in the traditional transportation planning framework [6]. Transportation planning organizations such as MPOs will be able to use this tool to include ITS in the transportation planning process and generate ITS options as an integral part of their TIP.

IDAS allows planners to evaluate ITS options as part of their overall transportation planning options. The software works with existing transportation planning software, such as EMME2, MinUTP, SYSTEM II, TRANPLAN, and TRANSCAD and uses planning data and output from existing transportation planning models to analyze and assess ITS benefits. A more complete description of IDAS can be found in Chapter 8.

U.S. DOT introduced IDAS as a near-term application to assess ITS benefits in the overall transportation network. An ongoing effort at U.S. DOT to develop a comprehensive transportation planning tool called *Transportation Analysis and Simulation System* (TRANSIMS) would eventually become a primary tool to integrate traditional transportation planning and ITS planning in a same suite of software.

6.6 Conclusions

Although ITS solutions and traditional approaches are different, the greatest benefits will be achieved when these are integrated. This process will start with integrating ITS projects into the transportation planning process. For ITS planning, the National ITS Architecture provides a wealth of resources. A small amount of currently available software packages could also aid in ITS planning.

Review Questions

1. What is the difference between traditional transportation and ITS planning?
2. What role can a metropolitan planning organization play in ITS planning?
3. What are the advantages of traceability-based ITS planning?
4. How could the National ITS Architecture be used as a resource in ITS planning?

5. How can ITS planning be integrated with the transportation planning process?

References

[1] Institute of Transportation Engineers, *Transportation Planning Handbook*, 2nd ed., Washington, D.C., 1999.

[2] U.S. Department of Transportation, *National ITS Architecture Implementation Strategy*, September 1998.

[3] Federal Highway Administration, *Integrating Intelligent Transportation Systems Within the Transportation Planning Process: An Interim Handbook*, Report No. FHWA-SA-98-048, January 1998.

[4] Federal Highway Administration, U.S. Department of Transportation, *Transportation Planning an ITS, Putting the Pieces Together* (prepared by Sarah J. Siwek and Associates), 1998.

[5] Virginia Department of Transportation, *Northern Virginia District Smart Travel Strategic Plan*, 2000.

[6] U.S. Department of Transportation, *ITS Deployment Analysis System*, 2001.

7
ITS Standards

One of the major motivations for the National ITS Architecture is to support the standard development activities. ITS standards support interoperability and interchangeability. The focus of this chapter is to discuss the ITS standard development process and its relationship with the National ITS Architecture. This chapter also presents the standards application areas and related ITS standards.

7.1 Introduction

One of the key objectives of ITS is to integrate a variety of previously independently operated components or systems to minimize redundancy and maximize efficiency. This integration requires that various components and systems from different vendors speak a common language so that they can understand each other. ITS standards are the means to establish a common language between different components or systems.

Many problems may arise from not implementing ITS standards. For example, an agency may have implemented a traffic management system using equipment from a manufacturer that has proprietary standards. When that agency wants to expand the system, it may only be able to do so by using equipment from the same manufacturer as the initial deployment because the same device or different devices offered by other manufacturers will not work on the same communications channel established in the initial deployment. With ITS standards, it will be possible to include any traffic signal controller from any vendor in a traffic management system.

ITS standards will support interoperability and interchangeability. Interoperability is achieved when many different types of devices could easily be used in the same communications channel. Interchangeability occurs when a device from different manufacturers can be used on the same communications channel [1].

7.2 Standard Development Process

ITS is a fusion of many different application areas. Consequently, the breadth of ITS standards encompasses many industries. Under the umbrella of the U.S. DOT's ITS Standards Program, a collection of representatives from the private sector, government, and other user groups have been actively involved in standards' development activities. Currently, ITS standards are created and published by the *Standard Development Organizations* (SDOs). Currently, the following organizations have been identified as SDOs:

- *American Association of State Highway and Transportation Officials* (AASHTO);
- *American National Standards Institute* (ANSI), Committee X12, Electronic Data Interchange;
- *American Society for Testing and Materials* (ASTM);
- *Electronics Industry Alliance/Consumer Electronics Association* (EIA/CEA)
- *Institute of Electrical and Electronic Engineers* (IEEE);
- *Institute of Transportation Engineers* (ITE);
- *National Electrical Manufacturers Association* (NEMA);
- *Society of Automotive Engineers* (SAE).

The standard development process starts with the development, approval, and publication of standards by SDOs. During this process, SDOs documents the technical details of standards. The standards then go through a voting process, where technical committees within the SDOs approve or disapprove the standards under review.

The next step is testing the approved standards. Under realistic operations conditions, these standards are tested to evaluate their functions, interoperability, interchangeability, and completeness. Once a standard passes the tests and matures, various vendors start providing different equipment with different functionalities conforming to the mature standards. The U.S. DOT then considers selected standards for adoption if they meet certain minimum criteria.

The Transportation Efficiency Act for the Twenty-first Century (TEA-21) specifies that any ITS project that receives federal funding should conform to the applicable or provisional standards and protocols. The standards considered by the U.S. DOT are first developed and approved by the SDOs.

The U.S. DOT is leading the standards development activities in the United States under its ITS Standards Program. Some of the primary objectives of program are to increase the public sector's option and ability to choose ITS products and services from multiple vendors and to facilitate interoperability at all levels.

7.3 National ITS Architecture and Standards

The National ITS Architecture provides a foundation for ITS standards activities. The logical and physical architecture define standards requirements, which serve as a foundation for standards development activities. Standards development activities are initiated by identifying which architecture flows (for the physical architecture) and data flows (from the logical architecture) are to be standardized.

Currently, the National ITS Architecture identifies which standards exist for a particular architecture flow. Once a physical architecture is developed and architecture flows between subsystems are identified, the National ITS Architecture could be consulted to see whether a standard exists for that particular architecture flow. Figure 7.1 shows how standards evolve from the National ITS Architecture [2]. Figure 7.2 shows representative architecture flows between the traffic management and transit management subsystems [2].

One can look at the National ITS Architecture documents to identify which standards have already been developed or are being developed for an

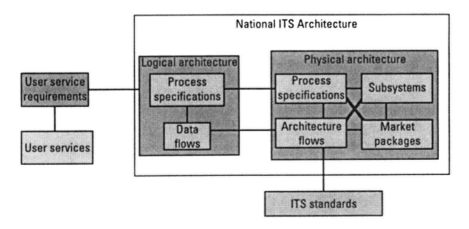

Figure 7.1 National ITS Architecture and standards. (*After:* [2].)

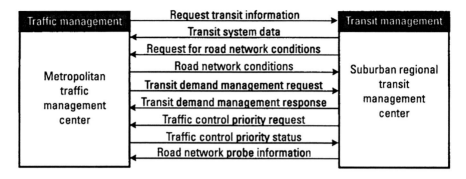

Figure 7.2 Architecture flow between subsystems. (*After:* [2].)

architecture flow. It is advised that the National ITS Architecture Web site be accessed to get the latest standards mappings to the the architecture flows.

7.4 ITS Standards Application Areas

One convenient way to present various ITS standards areas is by mapping these areas using the National ITS Architecture interfaces classes. These application areas by various National ITS Architecture interface class are shown in Table 7.1 [3]. Two types of standards—primary and secondary—are related to these application areas. Primary standards are the central standards that are required for the application area. Each application area may have several secondary standards, which are necessary to support the primary standards. Each primary and secondary standard may be associated with more than one application area.

7.4.1 Center-to-Roadside Interface Class

This section describes the center-to-roadside interface class.

7.4.1.1 Data Collection and Monitoring

This application area includes the interface between a traffic management center or a data archive and roadside equipment. Primarily, the interface is between the traffic management subsystem and roadway subsystem. The traffic management center remotely controls, monitors, and collects data from the equipment on or at the roadside. The roadside equipment collects and processes signals from the sensors as vehicles are detected to generate traffic information, such as speed and travel time. The roadside subsystem sends the information to the traffic management subsystem either by request or periodically through a prior plan. This

Table 7.1
ITS Standards Application Area Matrix

National ITS Architecture		
Interface Class	**Description**	**Standards Application**
Center-to-roadside	Standard for communications between a transportation Management center and roadway traffic control and data collection equipment	Data collection/monitoring
		Dynamic message signs
		Environmental monitor
		Ramp metering
		Traffic signals
		Vehicle sensors
		Video surveillance
Center-to-center	Standards for communications between transportation management centers	Data archival
		Incident management
		Rail coordination
		Traffic management
		Transit management
		Traveler information
Center-to-vehicle/ traveler	Standards for communications between transportation management centers and vehicle or traveler	Mayday
		Transit vehicle communications
		Traveler information
Roadside-to-vehicle	Standards for wireless communications between roadside equipment and vehicles on the road	Toll/fee collection
		Signal priority
Roadside-to-roadside	Standards for communications Between roadside equipment and railroad wayside equipment	Highway rail intersection

Souce: [3].

application area has several primary and secondary standards associated with it. The primary standards include the following:

- Object definitions for video switches (NTCIP 1208);
- Data dictionary for closed-circuit television (CCTV) (NTCIP 1205);

- Object definitions for environmental sensor stations and roadside weather information system (NTCIP 1204);
- Transportation system sensor objects (NTCIP 1209);
- Data collection and monitoring devices (NTCIP 1206).

7.4.1.2 Dynamic Message Signing

This application area includes the interface between a traffic management and a roadway subsystem. Dynamic message signing is the specific type of roadway subsystem that provides information such as traffic conditions, weather conditions, or any other travel advisory to motorists. This application area has one primary and several secondary standards. The primary standard is "object definitions for dynamic message signs" (NTCIP 1203).

7.4.1.3 Environmental Monitoring

This application area includes the interface between a traffic management and a roadway subsystem. The roadway subsystem includes an environmental sensor station that monitors air and water quality, emissions, and weather and roadway surface conditions. Various NTCIP standards support this application area. This application area has three primary and several secondary standards associated with it. The primary standards include the following:

- Object definitions for environmental sensor stations and roadside weather information system (NTCIP 1204);
- Data collection and monitoring devices (NTCIP 1206);
- Transportation system sensor objects (NTCIP 1209).

7.4.1.4 Ramp Metering

This application area provides an interface between a traffic management subsystem and a roadway subsystem. The roadway subsystem includes a ramp meter control unit, which controls the traffic in the freeway entry lanes to restrict entry. One primary and several secondary standards are associated with this application area. The primary standard is "ramp meter controller objects" (NTCIP 1207).

7.4.1.5 Traffic Signal

This application area provides an interface between a traffic management and roadway subsystem. The roadway subsystem includes a local signal controller or an on-street master controller that manages a group of local signal controllers. The traffic management center, based on the traffic situations data, identifies the appropriate signal timing plan (either pretimed or actuated) and sends this information to the local signal controller or the master controller, which implements the plan. This application area has two primary and several secondary standards. The primary standards include the following:

- Objects for signal systems master (NTCIP 1210);
- Object definitions for actuated traffic signal controller units (NTCIP 1202).

7.4.1.6 Vehicle Sensors

This application area includes the interface between a traffic management and roadway subsystem and a roadway and archived data management subsystem. The roadway subsystem includes roadway sensors that identify different traffic characteristics and communicates them back to a traffic management center. Roadway sensors also communicate data, through another interface, to agencies that use the data for archiving and off-line analysis. This application area has several primary and secondary standards. The primary standards include the following:

- Object definitions for video switches (NTCIP 1208);
- Data dictionary for CCTV (NTCIP 1205);
- Transportation system sensor objects (NTCIP 1209);
- Data collection and monitoring devices (NTCIP 1206).

7.4.1.7 Video Surveillance

This application area includes the interface between a traffic management and a roadway subsystem. The roadway subsystem includes video surveillance equipment, which includes CCTV cameras, video switches, and communications infrastructure that provides video images to the traffic management center. This application area has two primary and several secondary standards. The primary standards include the following:

- Object definitions for video switches (NTCIP 1208);
- Data dictionary for CCTV (NTCIP 1205).

7.4.2 Center-to-Center (C2C) Applications

This section presents the center-to-center applications.

7.4.2.1 Data Archival

This application area includes an interface between the archived data management subsystem and the sources and users of the archived data. The data archive collects data for off-line analysis purposes such as planning and research. Data sources for the archive include traffic management centers, transit management centers, emissions management systems, emergency management centers, and commercial vehicle administration systems. This application area is associated with two primary standards and numerous other secondary standards. The primary standards include the following:

- *Archived Data Management Subsystem* (ADMS) standards guidelines (ASTM AG);
- *ADMS data dictionary specifications* (ASTM DD).

7.4.2.2 Incident Management

This application area includes the interfaces between the emergency management subsystem and other related agencies. The primary purpose of the emergency management subsystem is to coordinate incident management activities such as detection, verification, response, motorist information, and traffic management. Currently, six primary standards and numerous supporting standards are associated with this application area. The primary standards include the following:

- Standard for common *incident management message sets* (IMMS) for use by *emergency management centers* (EMCs) (IEEE P1512-2002);
- Standard for traffic IMMS for use by EMCs (IEEE P1512.1);
- Standard for public safety IMMS for use by EMCs (IEEE P1512.2);
- Standard for hazardous material IMMS for use by EMCs (IEEE P1512.3);
- Standard for emergency management data dictionary (IEEE P1512.a);
- Message set for weather reports (NTCIP 1301).

7.4.2.3 Rail Coordination

This application area includes the interface between a traffic management center and a rail operations center. The traffic management subsystem sends information on the conditions at or at close proximity to a *highway-rail intersection* (HRI) that may impact the rail operations. Similarly, the rail operations center sends information on any maintenance activities or train operations that could affect the highway-rail intersection. There are currently no primary standards related to this application.

7.4.2.4 Traffic Management

This application area includes the interface between a traffic management subsystem and other centers, including transit management centers, information service providers, construction and maintenance, emergency management, toll administration, event promoter, media, and other traffic management centers. The interface enables data communications on real-time traffic and control data, transit and emergency operations, and construction and maintenance activities. There are three primary and numerous secondary standards for this application area. The primary standards include the following:

- Message sets for external TMC communications (ITE TM 2.01);
- Standard for functional level *traffic management data dictionary* (TMDD) (ITE TM 1.03);
- Message set for weather reports (NTCIP 1301).

7.4.2.5 Transit Management

This application area includes interfaces between a transit management center and other centers, including other transit management centers, information services providers, traffic management centers, financial institutions, enforcement agencies, and media and multimodal transportation service providers. The interfaces support various functions, including the coordination between transit agencies and other public transportation on scheduling, and between transit agencies and information service providers on incidents, schedules, and fares. Within NTCIP is the *Transit Communications Interface Profile* (TCIP) family of transit standards. Seven TCIP primary standards and several more secondary standards are part of this application area. The primary standards include the following:

- TCIP-*Control center* (CC) business area standard (NTCIP 1407);
- TCIP-*Common public transportation* (CPT) business area standard (NTCIP 1401);
- TCIP-*Fare collection* (FC) business area standard (NTCIP 1408);
- TCIP-*Incident management* (IM) business area standard (NTCIP 1402);
- TCIP-*Passenger information* (PI) business area standard (NTCIP 1403);
- TCIP-*Scheduling/runcutting* (SCH) business area standard (NTCIP 1404);

- TCIP-*Spatial representation* (SP) business area standard (NTCIP 1405).

7.4.2.6 Traveler Information

This application area includes the interfaces between the information service provider (ISP) subsystem, either a public or private sector creator of traveler information, and other centers that collect and/or disseminate traveler information. These interfaces support the roles of an ISP that may include information collection, intergration of collected data and dissemination of the aggregated data. This interface enables data communications on real-time traffic information, weather, transit, emergency operations, as well as construction and maintenance activities. There are four primary standards for this application area:

- Data Dictionary for Advanced Traveler Information System (ATIS) (SAE J2353);
- Message set for ATIS (SAE J2354);
- Messages for handling strings and look-up tables in ATIS standards (SAE J2540);
- Message set for weather reports (NTCIP 1301).

7.4.3 Center-to-Vehicle/Traveler

This section presents center to vehicle/traveler applications.

7.4.3.1 Mayday

This application area includes interfaces between a driver or traveler and emergency management and a transit management center. This interface enables a driver or traveler to either request emergency assistance or have such a request automatically sent following a crash. This application area includes only one primary standard, the onboard land vehicle mayday reporting interface (SAE J2313).

7.4.3.2 Transit Vehicle Communications

This application area includes interfaces between a transit management center and transit vehicles. Transit vehicles send information on location, passenger counts, maintenance, and so on to the transit management center. Similarly, the transit management center provides instruction on dispatch, routing, and other information. This application area includes several primary standards and few secondary standards. The primary standards include the following:

- TCIP-Control center (CC) business area standard (NTCIP 1407);

- TCIP-Common public transportation (CPT) business area standard (NTCIP 1401);
- TCIP-Fare collection (FC) business area standard (NTCIP 1408);
- TCIP-Incident management (IM) business area standard (NTCIP 1402);
- TCIP-Passenger information (PI) business area standard (NTCIP 1403);
- TCIP-Scheduling/runcutting (SCH) business area standard (NTCIP 1404);
- TCIP-Spatial representation (SP) business area standard (NTCIP 1405);
- TCIP-Onboard (OB) business area standard (NTCIP 1406).

7.4.3.3 Traveler Information

This application area includes interfaces between the centers that provide traveler information. The interfaces support information exchange for trip planning, route guidance, transit information, and so on. This application area includes six primary standards and numerous secondary standards. The primary standards include the following:

- Data dictionary for Advanced Traveler Information System (ATIS) (SAE J 2353);
- Message set for ATIS (SAE J 2354).
- Information Service Provider (ISP)–Vehicle location referencing standard (SAE J1746);
- Standard for ATIS message sets delivered over bandwidth restricted area (SAE J2369);
- Messages for handling strings and lookup tables in ATIS standards;
- Message set for weather reports (NTCIP 1301).

7.4.4 Roadside-to-Vehicle

This section presents roadside-to-vehicle applications.

7.4.4.1 Toll/Fee Collection

This application area includes interfaces between a toll collection or parking management facility and vehicles that would be paying the toll or fee. This interface supports reading vehicle and processing electronic identification and associated account information. The primary standards include the following:

- Standard specification for 5.9-GHz data link layer (ASTM TBD);
- Standard specification for 5.9-GHz physical layer (ASTM TBD);
- Specification for Dedicated Short Range Communication (DSRC) data link layer: medium access and logical link control (ASTM PS 105-99);
- Specification for Dedicated Short Range Communication (DSRC) physical layer using microwave in the 902–928-MHz (ASTM PS 111-98);
- Standard for message sets for vehicle/roadside communications (IEEE 1455-1999).

7.4.4.2 Signal Priority

This application area includes interfaces between traffic signal controllers and transit or emergency vehicles. The interfaces support providing priority to transit vehicles or preempting emergency vehicles, depending on the detection of the vehicle type or requests from vehicles. Five primary and several supporting standards are associated with this application area. The primary standards include the following:

- Standard specification for 5.9-GHz data link layer (ASTM TBD);
- Standard specification for 5.9-GHz physical layer (ASTM TBD);
- Specification for Dedicated Short Range Communication (DSRC) data link layer: medium access and logical link control (ASTM PS 105-99);
- Specification for Dedicated Short Range Communication (DSRC) physical layer using mirowave in the 902–928-MHz (ASTM PS 111-98);
- Objects for signal control priority (NTCIP 1211).

7.4.5 Roadside-to-Roadside

This section presents roadside-to-roadside applications.

Highway-Rail Intersection

This application area includes interfaces between railway and roadside equipment. The interfaces support coordinated operations of the railway- and roadway-side equipment to improve operations and safety for both rail transit and highway vehicles. This application area includes one primary standard, the standard for interface between the rail subsystem and the highway subsystem at a highway rail intersection (IEEE P1570).

As can be seen, numerous ITS standards have been developed to support different application areas of ITS. Among these standards, the NTCIP is the largest family. Section 7.5 discusses this standard in detail.

7.5 NTCIP

NTCIP consists of a group of protocols and data definitions to support the communications between a transportation management center and roadway equipment and between transportation management centers [1]. NTCIP also contains transit domain standards TCIP. One of the major motivations in the development of NTCIP was to build on accepted industry-standard Internet communications protocol. NTCIP standards are being developed by several SDOs, namely NEMA, AASHTO, and ITE.

One of the primarily objectives of the National ITS Architecture is to promote interoperability through standards development. The National ITS Architecture defines the interface requirements between various systems or devices. ITS standards, including NTCIP, are being developed from those interface requirements. Figure 7.3 depicts NTCIP's role in the National ITS Architecture [1]. As shown in the figure, the center-to-center and center-to-field communications between various ITS physical elements are supported by the NTCIP protocols. Figure 7.4 depicts the myriad of NTCIP center-to-field and center-to-center interfaces that are supported by the NTCIP C2C protocols currently being developed [1].

NTCIP is currently associated with the following ITS application areas:

1. Center-to-roadside:
   - Data collection and monitoring;
   - Dynamic message signing;
   - Environmental monitoring;
   - Ramp metering;
   - Traffic signaling;
   - Vehicle sensing;
   - Video surveillance.
2. Center-to-center:
   - Data archival;
   - Incident management;
   - Traffic management;
   - Transit management;
   - Traveler information.

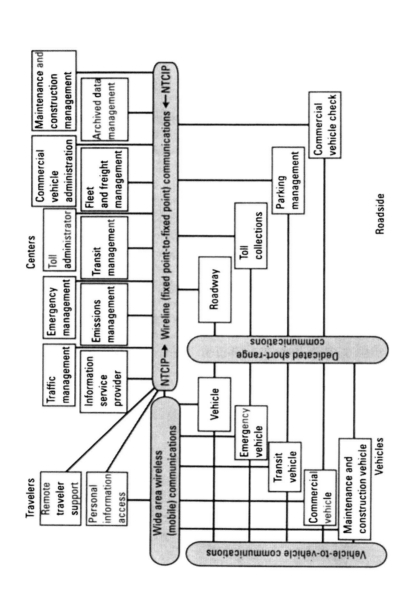

Figure 7.3 NTCIP in the National ITS Architecture. (*From:* [1]. Reprinted from NTCIP 9001, V03.02, October 2002, The NTCIP Guide, by permission of NEMA. Excerpt @ 2002 AASHTO, ITE, NEMA.)

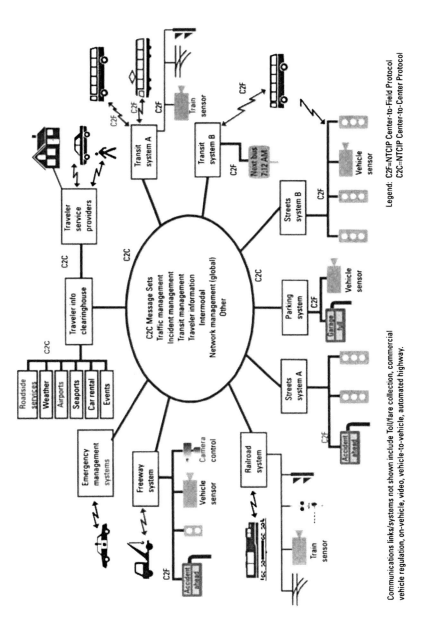

Figure 7.4 NTCIP in diverse ITS areas. (*From:* [1]. Reprinted from NTCIP 9001, V03.02, October 2002, The NTCIP Guide, by permission of NEMA. Excerpt @ 2002 AASHTO, ITE, NEMA.)

3. Center-to-vehicle/traveler:
   - Transit vehicle communications;
   - Traveler information.
4. Roadside-to-vehicle:
   - Signal priority.

7.6 Standards Testing

Standards are tested to ensure that they serve the intended purpose of performance, reliability, interoperability, functionality, and interchangeability [4]. By and large, the ITS Standards Program has not required that the various ITS standards activities document the systems engineering artifacts necessary for thorough testing. There is an effort underway to correct this deficiency and allow for the consistent testing of ITS standards. During the development process, standards should be continually tested to ensure that they satisfy all the requirements set forth at the start of the process. This is called validation testing.

Another type of standards testing is verification testing, which is conducted by the vendors or users. They examine the practicality and economic viability of building systems that conform to the standards. A verification test could be performed by reviewing and analyzing the standards documents or developing software or hardware based on the standard. A third type of testing is experienced-based testing, which includes real-world experience with systems built based on standards.

Standards testing provides useful information to users. Testing efforts help standards developers modify any standards as required. In addition, testing will help build user confidence in ITS equipment and services. The U.S. DOT has initiated a national standards testing program in the United States and has assigned an independent company the task of leading the testing.

7.7 Conclusions

There has been a considerable amount of success and progress from the ITS Standards Program. There is a large breadth of ITS standardization currently underway in the United States. However, many of the ITS standards activities have focused on standardization of their particular area without ensuring compatibility with ohter ITS standards activities. Extreme care should be exercised with regard to specifying and deploying ITS standards. A motivation for being an early adopter of ITS standards is that TEA-21 requires that any Highway Trust Fund–supported project should conform to the National ITS Architecture and applicable the U.S. DOT–adopted standards. Even though the

U.S. DOT has not adopted any standards at the time of this writing, there can be a great future value to analyzing and incorprating the existing ITS standards in deployment for the good of the ITS industry as a whole.

Ongoing activities on ITS standards can be found on the Internet at http://www.its-standards.net and from various sources, such as the U.S. DOT and the SDOs. Additionally, ongoing and planned activities for the NTCIP family of standards can be found on the NTCIP Web site: http://www.ntcip.org.

Review Questions

1. What is an ITS standard? What is the purpose of ITS standards?
2. Describe a typical standard development process.
3. What is the relationship between the National ITS Architecture and standards? How is standards development supported by the National ITS Architecture?
4. What is NTCIP? What types of ITS applications are supported by the NTCIP? How is NTCIP related to the National ITS Architecture?
5. Why is ITS standards testing important?

References

[1] AASHTO, ITE, NEMA, NTCIP Guide, NTCIP 9001, V0302, October 2002.

[2] U.S. Department of Transportation, National ITS Architecture, updated 2002.

[3] http://www.its-standards.net, U.S. DOT's standards Web site, 2003.

[4] U.S. DOT, ITS Standards Testing Program, "How Do We Know That the Standards Work?" Washington, D.C., June 12, 2001.

8

ITS Evaluation

8.1 Introduction

One of the most challenging aspects of ITS is that the technology that serves its various functions changes very rapidly. Evaluation becomes important with each cycle of potential new and deployed technology. There are many advantages to ITS evaluation, one of which is that it minimizes the risk of project failure through unmet objectives.

Evaluation is an integral part of any system development process. For example, as shown in Figure 8.1, evaluation is an integral part of the life cycle of an ITS project. There are three types of evaluation: predeployment evaluation (at the planning level), deployment tracking, and impact assessment. Evaluation can start as early as the planning stage, where competing projects are prioritized and selected based on available resources for a fiscal year. As shown in the figure, once the project is operational, a system assessment or evaluation will be required. This assessment helps identify whether the system is performing as envisioned during the planning stage.

Two types of evaluation could be conducted after the project is operational. The first is an inventory of the system that is deployed. This type of evaluation is also called "deployment tracking" and assesses the actual deployment level in relation to the plan [1]. The other type of evaluation is an assessment of the changes in impacts, such as travel time and crash rates, of selected attributes in relation to the predeployment conditions.

ITS is sometime an unfamiliar concept to many people, including decision makers and implementers who are reluctant to accept ITS because they are unsure whether the investment would be cost effective. In this case, evaluation results of similar applications would have satisfied future uncertainty regarding

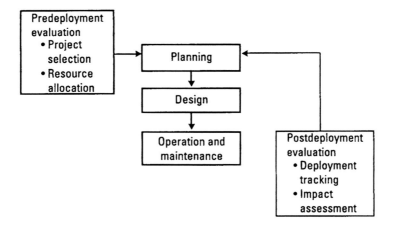

Figure 8.1 Role of evaluation in ITS projects.

ITS applications. In the absence of first-hand experience with the application, these types of evaluation results would be the best resource available to aid decision makers on future investments in ITS.

As stated earlier, three types of evaluation can be conducted: at the planning level; in deployment tracking after the projects are partially or fully operational; and an assessment of the project's impact based on selected measures of effectiveness after the project is operational. These are discussed next.

8.2 Project Selection at the Planning Level

Data from previous studies could be used to assess the impacts of a planned system. Computer programs such as IDAS could be used to compute the impacts of a planned system. The following two methods could be applied when prioritizing competing projects.

8.2.1 Benefit-Cost Analysis

Benefit-cost analysis has been widely used by transportation agencies to prioritize and justify projects. In benefit-cost analysis, benefits are identified and converted into monetary units to make them divisible by cost. Usually, the ratio of benefit to cost is computed, where the benefit represents the current worth of the benefits taken over the life of the project, and the cost represents the current worth of initial capital and future costs.

8.2.2 Relative Rating and Ranking

Some agencies use a form of the weighting-based method to select between alternative projects, because this method does not require that the evaluation criteria be converted into monetary units. This method is represented by the following equation [2]:

$$S_i = \sum_{j=1}^{N} K_j V_{ij} \qquad (8.1)$$

where
S_i = total value of score of alternative i;
K_j = weight placed on criteria j;
V_{ij} = relative weight achieved by criteria j for alternative i.
The alternative with the highest total score is selected for implementation.

8.3 Deployment Tracking

The Federal Highway Administration tracks the level of ITS deployment in the United States to ensure that we succeed in meeting goals and determine where gaps exist between goals and actual implementation. On a state and regional level, a survey of statewide or regional transportation agencies usually provides data that are used to evaluate deployments in relation to the initial objectives. This exercise also helps identify, at the current progress rate, what level of future deployment could be expected and can also help determine what actions need to be taken to achieve a desired level of deployment by a target time frame.

The U.S. DOT has made significant efforts to track ITS deployment as part of an effort to evaluate the achievement of national goals in this area. The initial survey of transportation agencies throughout the United States concentrated on major metropolitan areas. Figure 8.2 shows the survey used to track data for electronic toll collection. Similarly, different sets of survey questions were developed for other areas, such as arterial management, transit management, freeway management, and emergency management. A set of questions was also developed for metropolitan planning organizations. Survey questions related to these areas and additional information on the national deployment tracking effort can be found at http://itsdeployment2.ed.ornl.gov/its2000/ default.asp.

One of the major indicators of the level of progress in ITS deployment is integration. Integration is defined as the communication of information or data between agencies associated with various ITS infrastructure components, such as freeway surveillance systems and transit automatic vehicle location systems. In metropolitan areas, these agencies are identified as the following:

2000 Electronic Toll Collection Survey — Metropolitan area goes here

Responses for the questions should be limited to the boundary depicted in the enclosed map of your metropolitan area and not for the entire toll system operated by your agency. (NR = No Response)

1. Please review the first two columns containing information your agency provided in 1999. Enter any changes for 2000 in the last two columns or check the box at the bottom if there are no changes.

	Total in 1999	1999 Estimated Total by 2005	Total in 2000	2000 Estimated Total by 2005
Total number of toll collection plazas operated	☐	☐	☐	☐
Total number of toll collection plazas with Electronic Toll Collection (ETC) capabilities	☐	☐	☐	☐
Total number of toll collection lanes	☐	☐	☐	☐
Total number of toll collection lanes with Electronic Toll Collection (ETC) capabilities	☐	☐	☐	☐

☐ **Verified 1999 data; no change**

	1999 Response	2000 Response
2. Your tags are used by other toll operators in your metropolitan area	☐	☐

Please indicate the agency

☐ Local agencies go here
Other (please specify)________________
Other (please specify)________________

	1999 Response	2000 Response
3. Your toll tags are used by operators of public transit to pay fares in your metropolitan area	☐	☐

Please indicate the agency

☐ Local agencies go here
Other (please specify)________________
Other (please specify)________________

Agency name goes here 1

Figure 8.2 Survey questionnaire used to track electronic toll collections. (*After:* [1].)

- State DOTs responsible for freeway and incident management components;
- Local governments responsible for arterial management components;
- Public transit agencies responsible for transit management components.

ITS Evaluation

```
2000 Electronic Toll Collection Survey                    Metropolitan area goes here

QUESTIONS 4 THROUGH 10 RELATE TO THE NATIONAL ITS ARCHITECTURE.

4. Have any members of your staff attended USDOT-sponsored National ITS Architecture training
courses?

   ☐ Yes
   ☐ No
         If not, why not? (check as many as apply)

              ☐ We are unaware of the availability USDOT-sponsored architecture training courses.
              ☐ We are aware of such training, but have no funding to support staff participation.
              ☐ We plan to send a member of our staff to the training courses within the year.
              ☐ We are aware of such training, but it is not a priority to send staff to participate.
              ☐ Other (please specify)_____

   ☐ Don't Know

5. Is your agency involved in an organized effort to develop a regional ITS architecture?

   ☐ Yes
         If yes, what is the status of the regional architecture?
              ☐ Our region has a fully developed regional ITS architecture undergoing continuing
                development and updating.
              ☐ Our regional ITS architecture is under initial development
   ☐ No
         If not, why not?
              ☐ There is no such effort underway in our region. Go to question 10
              ☐ There is such an effort in our region, but we are not involved with it. Go to question 10

   ☐ Don't know if my agency is involved with architecture development. Go to question 10

6. If you answered yes to question 5, what other agencies are involved, and which one is the lead for
the effort? Check the type(s) of agency involved in the effort (Do not check your own agency type unless there
is another agency of your type that is involved with the effort.) Circle the agency that is leading the regional
architecture effort (If you are the lead agency, circle your agency type; if it is also checked, we will know that you
are the lead and there is another agency of your type involved in the effort.)

         ☐ State department of transportation
         ☐ County highway authority(s)
         ☐ City transportation department(s)
         ☐ Transit property(s)
         ☐ Rail agency
         ☐ MPO
         ☐ Fire department(s)
         ☐ Local police department(s)
         ☐ State police/Highway patrol
         ☐ Other emergency services provider(s) (please specify)_____
         ☐ Toll authority(s)
         ☐ Airport authority
         ☐ Other port authority (please specify)_____
         ☐ Freight shippers (private sector)
         ☐ Traveler information service providers (private sector)
         ☐ Other (please specify)_____
         ☐ Don't know

Agency name goes here                    2
```

Figure 8.2 Continued.

The U.S. DOT defined the level of integration as high, low, and medium. A metropolitan area has a high level of integration when the three agencies listed in the bullet points are communicating (though integration may not be complete). A medium level of integration exists when two out of the three agencies

2000 Electronic Toll Collection Survey

Metropolitan area goes here

7. What is the nature of the regional architecture?
- ☐ Encompasses a single county
- ☐ Encompasses more than one county
- ☐ Encompasses entire state
- ☐ Encompasses a corridor
- ☐ Don't know

8. Have you attempted to develop project architectures within your regional architecture? If so, how many?
- ☐ Yes Number: _____
- ☐ No
- ☐ Don't know

9. How long has your agency been involved with the region's architecture development effort?
- ☐ Less than one year
- ☐ One to two years
- ☐ Longer than two years
- ☐ Don't know

10. Has any organization provided you with information concerning architecture development activities?
- ☐ No — Go to question 11
- ☐ Don't know — Go to question 11
- ☐ Yes — Please indicate which organization and check if the information was useful.

Organization	Check if Information Was Received	Check if Information Was Useful
FHWA	☐	☐
ITS America State Chapter	☐	☐
FTA	☐	☐
ITS America (national)	☐	☐
APTA	☐	☐
ITE	☐	☐
AASHTO	☐	☐
APA	☐	☐
AMPO	☐	☐
Other (please specify) _____	☐	☐

11. Is your agency willing to share cost information on ITS-related equipment (i.e., capital and O&M cost, and brief equipment description)? This information will be used to update the ITS JPO sponsored ITS unit cost database. This database provides ITS cost data for ITS implementation and is accessible at the following URL: http://www.its.dot.gov/eval/itsbenefits.htm
- ☐ No
- ☐ Yes, please provide name and phone number of the cost information contact if different from respondent. This person will be contacted for the cost information at a later date.

Agency name goes here

Figure 8.2 Continued.

are communicating. A metropolitan area is said to have a low level of integration when none of the agencies are communicating.

Figure 8.3 shows the progress in integrated metropolitan ITS deployment for 1997, 1999, and 2000 at high, medium, and low levels [1]. As shown in the

> **2000 Electronic Toll Collection Survey**   Metropolitan area goes here
>
> If there is anything else you want to tell us about any ITS efforts in your agency, please use this space for that purpose. Also, any comments you wish to make that you think may help us in future efforts to track ITS deployment will be appreciated, either here or in a separate letter.
>
>
>
> Your contribution to this effort is greatly appreciated. If you would like to receive a copy of your metropolitan area report and the national summary report, please indicate below. (Circle the number of your answer)
>
> ☐ Yes, send a copy of the reports to me.
> ☐ No, do not send a copy of the reports to me.
>
> Agency name goes here   4

Figure 8.2 Continued.

figure, from 1997 to 2000, the distribution of metropolitan areas by low, medium, and high changed from 39, 25, and 11 to 23, 28, and 24, respectively. As the figure suggests, more and more metropolitan areas are adopting ITS and trying to maximize the benefits by integrating components between agencies.

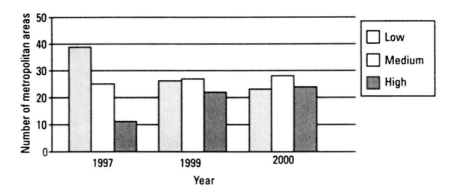

Figure 8.3 Progress in integrated metropolitan ITS deployment (for each year, the first, second, and third columns represent low, medium, and high, respectively). (*After:* [1].)

8.4 Impact Assessment

ITS impacts have usually been assessed in various physical contexts, such as crashes, traffic flow, air quality, and fuel consumption. Although economic impacts, such as employment, income, and effects on property, and social impacts, such as accessibility of facility and services, have usually not been considered, many projects are influenced by these factors. As ITS is mainstreamed or becomes an integral part of our transportation systems, these economic and social factors will become part of the evaluation and decision-making criteria. The following section discusses the physical impacts of ITS.

Table 8.1 shows the most commonly used performance criteria and corresponding measures of effectiveness (referred to as "few good measures" for ITS evaluation). The selection of performance measures depends on the relative importance to the users and the resources available for evaluation.

Safety impacts could be measured by estimating the changes in crash, injury, or fatality rates between the before condition (before ITS deployment) and after condition (after ITS deployment). Police crash reports are the primary source of this data. Safety impacts can also be estimated by traffic conflict analysis before and after ITS implementation. Video images of before and after conditions could provide data for conflict analysis.

Travel time could be measured either in the field or by computer simulation. In the field, the "floating car" or "average speed" method could be used to compute travel time. In the floating car method, the driver of the test vehicle tries to drive at average speed of the traffic by passing as many vehicles as that pass the test vehicle in the corridor being evaluated. In the average speed method, the driver of the test car drives the test section at a representative speed (as perceived by the test car driver) of the traffic. The following equation can be used to determine the sample size or the minimum number of test runs required [2]:

Table 8.1
Measures of Effectiveness by Performance Criteria

Performance Criteria	Measure of Effectiveness
Safety	Changes in: • Crashes • Injuries • Fatalities
Travel time	Travel time for selected origin-destination Overall systemwide reduction in travel time Reduction in delays by mode
Throughput	Vehicle using facility (links, segments) Persons using facility (links, segments)
Customer satisfaction	Mode change Customer ratings of travel experience
Air quality/emission	Reduction in: • Carbon monoxide • Nitrous oxide • Ozone • Volatile organic compounds • Hydrocarbon
Fuel consumption	Reduction in fuel consumption by users

$$N = \left(t_\alpha \times \sigma/d\right)^2 \qquad (8.2)$$

where

N = sample size;
σ = standard deviation;
t_α = value of student's t distribution with confidence level of $(1 - \alpha/2)$; and degrees of freedom of $(N-1)$;
d = limit of permitted error in the speed estimate;
α = significance level.

The limit of permitted error, d, depends on the study type. The following is the suggested ranges of d based on study type [3]:

- Transportation planning and highway needs studies: ±3.0 to ±5.0 mph;

- Traffic operation, trend analysis, and economic evaluations: ±2.0 to ±4.0 mph;

- Before-and-after studies: ±1.0 to ±3.0 mph.

Throughput or flow could be estimated by field data on the number of vehicles or people passing a specific point in the study segment in a time period, which could be taken as an hour. Before-and-after data could be used to estimate any changes in the processing capability of the study section.

Customer satisfaction could be measured by identifying how many people change their choice of mode or rating of travel experiences from before ITS deployment to after ITS deployment. Survey questionnaires sent by mail to travelers in the study area or an on-site customer survey could be used to qualitatively assess changes in customer satisfaction.

Changes in air quality from the before to after condition could be measured by changes in the level of harmful gases, such as carbon monoxide, hydrocarbon, and ozone, in the environment. Equipment could be used in the field to collect data on the level of these gases in the air of the study area during the before and after periods to assess any changes. Another relatively less resource-intensive method includes the use of simulation programs to estimate changes in the level of these harmful gases in the environment.

Changes in fuel consumption could be estimated as a surrogate of travel time or delay reduction. Simulation programs could also be used for this purpose.

8.5 Benefits by ITS Components

Any realistic benefit estimation should be systematic and logical. The estimation approach may differ depending on the ITS component and corresponding elements or program areas that are subject to evaluation. Table 8.2 shows various ITS components and corresponding elements.

Since the inception of the ITS concept in the early 1990s, many studies have been completed that report its various benefits. Most of these studies reported positive impacts of ITS deployments. Figure 8.4 summarizes the reported benefits of ITS deployments in metropolitan areas. Similar benefits data could also be found for other ITS components.

8.6 Benefit Estimation Categories

The Federal Highway Administration describes three benefit categories: measured, anecdotal, and predicted [4]. Figure 8.5 show these benefit categories and associated general analysis tools. Measured benefits are directly based on field data and considered the most reliable. For example, the measured travel time benefits of implementing a freeway management system in an area will be based on data collected through a probe, roadway sensors, or video surveillance system. Anecdotal benefits of a project are based on estimates of the people directly

Table 8.2
ITS Components and Elements

Components	ITS Element
Metropolitan ITS infrastructure	Arterial management systems
	Freeway management systems
	Transit management systems
	Incident management systems
	Emergency management
	Electronic toll collection
	Electronic fare payment
	Highway-rail intersections
	Regional multimodal traveler information
Rural ITS infrastructure	Crash prevention and security
	Emergency services
	Travel and tourism
	Traffic management
	Transit and mobility
	Operations and maintenance
Commercial vehicle operations	Safety assurance
	Credentials administration
	Electronic screening
Intelligent vehicles	Driver assistance
	Collision avoidance and warning

Source: [5].

associated with it. This benefits assessment is perceived as less reliable because it may be biased in certain cases.

Finally, the predicted benefits are estimated through some type of predictive analysis or simulation. Benefits are predicted when field data are not available or data collection is not economical. Adaptation of any particular method for benefits measurement depends on the available resources and needs for the level of estimates (i.e., the level of reliability).

8.7 Evaluation Guidelines

It may become rather costly to collect field data periodically to evaluate system deployment. ITS proponents agree that ITS will not be mainstreamed until its

Metropolitan Benefits By Program Area		
Program Area/Benefit Measure		Summary
Arterial Management Systems	Safety Improvements	Automated enforcement of traffic signals has reduced violations 20% to 75%
	Delay Savings	Adaptive Signal Control has reduced delay from 14% to 44%
	Throughput	
	Customer Satisfaction	72% of surveyed drivers felt "better off" after signal control improvements in Michigan
	Cost Savings	Transit Signal Priority on Toronto transit line allowed same service with one less vehicle
	Environmental	Improvements to traffic signal control have reduced fuel consumption 2% to 13%
	Other	Adaptive Control has reduced stops from 10% to 41%
Freeway Management Systems	Safety Improvements	Ramp Metering has shown 15% to 50% reduction in crashes
	Delay Savings	11 to 93.1 vehicle hours reduced due to ramp metering I-494, Minneapolis
	Throughput	Systemwide study in Minneapolis - St. Paul found 16.3% increase in throughput
	Customer Satisfaction	After Twin Cities shutdown, 69% of surveyed travelers support modified continued operation
	Cost Savings	Georgia Navigator $44.6 million/year in incident delay reduction (integrated system)
	Environmental	
	Other	Ramp Metering has shown 8% to 60% increases in speed on freeways
Transit Management Systems	Safety Improvements	AVL with silent alarm supported 33% reduction in passenger assaults on Denver System
	Delay Savings	Reported improvements in on-time performance from 9% to 23% with CAD/AVL
	Throughput	
	Customer Satisfaction	Customer complaints decreased 26% after Denver installed CAD/AVL
	Cost Savings	AVL reduced San Jose paratransit expenses from $4.88 to $3.72 per passenger
	Environmental	
	Other	Reductions in fleet size from 4% to 9% due to more efficient bus utilization
Incident Management Systems	Safety Improvements	San Antonio, TX reports reduced crash rate of 41%
	Delay Savings	Reductions range from 95 thousand to 2 million hours per year
	Throughput	
	Customer Satisfaction	Customers very satisfied with service patrols (hundreds of letters)
	Cost Savings	Cost Savings from $1 to $45 million per year, varying with extent of system
	Environmental	TransGuide reduced fuel consumption up to 2600 gal/major incident
	Other	
Emergency Management Systems	Safety Improvements	
	Delay Savings	
	Throughput	
	Customer Satisfaction	95% of drivers equipped with PushMe Mayday system felt more secure
	Cost Savings	
	Environmental	
	Other	
Electronic Toll Collection	Safety Improvements	Carquinez Bridge, CA: Increase in crashes (27 to 30) and injuries between 1996 and 1997
	Delay Savings	Carquinez Bridge, CA: person time savings of 79,919 hours (per year) or about $1.07 million
	Throughput	Tappan Zee Bridge: Manual lane 400-450 vph, ETC lane 1000 vph
	Customer Satisfaction	
	Cost Savings	Roadway Maintenance can be reduced 14%
	Environmental	Florida: Reduced CO 7.3%, HC 7.2%, Increased NO$_x$ 34% with 40% ETC usage
	Other	Value pricing using ETC in Florida resulted in 20% of travelers adjusting departure time
Electronic Fare Payment	Safety Improvements	
	Delay Savings	
	Throughput	
	Customer Satisfaction	In Europe, 71% to 87% user acceptance of coordinated smart cards for transit/city services
	Cost Savings	New Jersey Transit estimates $2.7 million cash handling reduction annually
	Environmental	
	Other	
Highway-Rail Intersections	Safety Improvements	92% of train engineers felt safety equal or greater with automated horn warning system
	Delay Savings	
	Throughput	
	Customer Satisfaction	School bus drivers felt in-vehicle warning devices enhanced awareness of crossings
	Cost Savings	
	Environmental	Automated horn warning system reduced noise impact area by 97%
	Other	
Regional Multimodal Traveler Information	Safety Improvements	Crash rate for drivers using web traveler information in San Antonio reduced 0.5%
	Delay Savings	San Antonio modeling results indicate a 5.4% reduction in delay for web site users
	Throughput	
	Customer Satisfaction	38% of TravTek Users found in-vehicle navigation useful in unfamiliar areas
	Cost Savings	ROUTES (London): estimated 1.3 million pounds sterling due to increased transit ridership
	Environmental	SmartTraveler Boston: estimated reductions NO$_x$ 1.5%, CO 33%
	Other	

Source: www.benefitscost.its.dot.gov   *Database also includes negative impacts of ITS.   Date: 3/14/2001
Table ES-2: Benefits Database Desk Reference

Figure 8.4 Reported benefits in metropolitan ITS infrastructure. (*After:* [5].)

benefits become transparent to the users and decision makers. With its vision to mainstream ITS, the U.S. DOT led many major evaluation efforts in the United States.

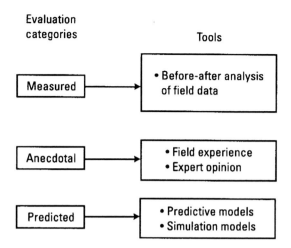

Figure 8.5 Evaluation categories and tools.

The TEA-21 provided guidelines for evaluating ITS projects. These guidelines apply to any operational test and deployment project and include requirements for objectivity and independence of the project evaluators so an unbiased evaluation can be done while still providing adequate funding for evaluation. More information on TEA-21 guidelines for ITS evaluation can be found on the Web site http://www.its.dot.gov/eval/evalguidelines_tea21evalguidelines.htm.

8.8 Evaluation Support Tools

Several different software programs could be used in ITS evaluation. These tools could provide an efficient means to evaluate an ITS project, given the limited budgets available for such tasks.

8.8.1 ITS Deployment Analysis System

Evaluators can now use software to support their evaluation efforts. Some of these software programs directly compute benefits and costs based on required data input. Some programs provide the impacts of various ITS deployments in terms of travel time, speed, delay, and so on, which could be used as inputs in a cost and benefit study.

The U.S. DOT developed the IDAS to support the evaluation of alternative ITS deployments. IDAS supports the evaluation of ITS projects at the project, component, and equipment levels. This software program is now commercially available. Knowledge of travel-demand forecasting, ITS, and travel-demand model files (node coordinates, network links, and trip tables) is necessary to use IDAS.

Figure 8.6 shows various components of the IDAS model and the interfaces between them. As shown in the figure, IDAS has three distinct modules: benefits, cost, and alternative comparison. IDAS can be interfaced with current planning models where it uses output from regional transportation planning models as input. With these input variables, the program estimates the impacts of ITS in terms of various measures of effectiveness such as cost, safety, emissions, and travel time.

8.8.2 Traffic Simulation Models

Traffic simulation models such as INTEGRATION, DYNASMART, and DYNAMIT are suitable for evaluating the operational impacts of ITS projects. Simulation models help provide a cost-effective analysis of the deployed system. Earlier simulation models such as CORSIM did not provide opportunities to

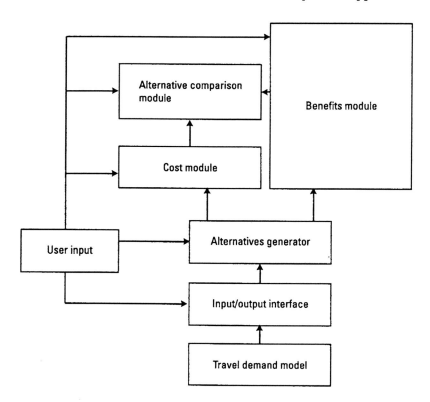

Figure 8.6 Modules and interfaces of IDAS. (*After:* [6].)

model ITS exclusively, such as allowing users to model variable message signs or ramp metering, although CORSIM does allow modeling adaptive traffic signal control. Unlike CORSIM, these software programs were developed to explicitly model ITS deployments. More discussions of some of these simulation models are presented in Chapter 2.

8.9 Challenges for ITS Evaluation

Institutional issues, not technical items, are often the major obstacle to evaluation. Coordination between the project partners and the project-partners and evaluators is the key to supporting evalution tasks, such as selection of the performance criteria and the collection of correspoindng data.

One of the major challenges for ITS evaluation is sorting out the ITS impacts when ITS was deployed as part of a larger improvement project. How do we sort out the benefits that exclusively came from ITS versus those from other project components? Statistical analysis that can exclusively account for the effects of ITS impacts could be used to address this problem.

Another major challenge is the availability of the required data for evaluation. In addition, resource requirements for data collection may also become a constraint with the limited budgets dedicated for evaluation. If an evaluation plan is in place and data requirements are identified ahead of time, then the responsible agencies could periodically collect field data. Coordination among stakeholder agencies in data collection and analysis will also help meet some of these challenges.

One study reported the challenges of data availability in the evaluation of automated enforcement programs due to a lack of before-and-after crash and violation data [7]. This study used a statistical technique called meta-analysis to estimate the effect of red light running cameras and violations at signalized intersections. This statistical technique is based on summarizing the results of previous studies and developing an estimate of the average effect of a deployed measure. When lacking the required data for direct analysis, this type of statistical tool could be applied to estimate the impacts of ITS measures.

Questions may also be raised regarding the interpretation of the impact data. For example, how does an agency use the data from a study that reported that implementing ramp metering has been found to reduce delay by a certain percentage? Several exposure conditions at the study site may differ from a region that may want to use the data. The agency needs to know how these results will transfer to other regions and how the data can be calibrated for a particular region. The greatest number of benefits will be achieved when these impact data also describe how to apply or calibrate these data for a different region.

Collaboration between the evaluators and project personnel as early as possible, ideally at the start of the project, would allow proper design intergration into the project to collect data needed for evaluation, but not necessarily needed for operation. This would allow collection of the required data for evaluation.

8.10 Conclusions

Periodic assessment is important to keeping ITS programs on an optimal and orderly track. Transportation agencies should maintain provisions for evaluating the technical as well as financial processes. In addition, a thorough evaluation during the project selection process also supports improved returns on investment.

Evaluation is a continuing engagement that provides "lessons learned" to avoid past mistakes. Transportation agencies can also realize the benefits of evaluation not only for keeping the ITS programs on track, but also for marketing the programs to policy makers and the public.

An agency could obtain information to evaluate a potential or existing ITS project from a number of on-line databases. In addition, many software programs, specifically those developed for ITS, are excellent resources for justifying or prioritizing ITS projects.

Although ITS evaluations pose many challenges, such as cost, data availability, appropriate methods, and interpretation and application of the evaluation data, as described in Section 8.9, a review of previous practices, ongoing efforts, and future studies will lead to opportunities for better evaluations of ITS projects. As ITS becomes an integral part of transportation projects such as construction and maintenance, more challenges may arise from differentiating the impacts of ITS from the overall improvement project. Public agencies will look for tools that address the evaluation challenges for ITS projects.

Evaluation must be incorporated into the life cycle planning, so resources and time can be allocated during the planning stage. An unbiased evaluation of an ITS project is an essential criteria for a successful program that mainstreams ITS. Some projects may take many years to be fully deployed, so an interim evaluation will be needed to measure the long-term viability of a project and evaluate whether there is a need to change the original plan. Therefore, interim evaluations may be necessary as a predefined milestone in the life cycle to gauge the progress toward goals and objectives.

Review Questions

1. What is deployment tracking? Why is it considered a part of ITS evaluation?
2. What are some of the impacts of ITS? How could they be measured?
3. Perform a study on ITS in your area. Based on published results, estimate their impacts.
4. What are some of the challenges of ITS evaluation? How could they be overcome?
5. Describe three ITS evaluation support tools, including where they could be applied.
6. Why is ITS evaluation important? What is the role of ITS evaluation in the future of our transportation system?

References

[1] U.S. Department of Transportation, "ITS Deployment Tracking," http://www.itsdeployment2.ed.ornl.gov/its2000/, 2003.

[2] Garber, N., and L. Hoel, *Traffic and Highway Engineering*, 3rd ed., Pacific Grove, CA: Brooks/Cole, 2002.

[3] Institute of Transportation Engineers, *Traffic Engineering Handbook*, 4th ed., Washington, D.C., 1992.

[4] Federal Highway Administration, *ITS Benefits: Continuing Successes and Operational Test Results*, Washington, D.C., October 1997.

[5] Federal Highway Administration, *Intelligent Transportation Systems Benefits: 2001 Update*, Report No. FHWA-OP-01-024, June 2001.

[6] Federal Highway Administration, "ITS Deployment Analysis System," *ITS America Annual Meeting*, Boston, MA, May 2000.

[7] Flannery, A., and R. Maccubbin, "Using Meta-Analysis Techniques to Assess the Safety Effect of Red Light Running Cameras," Mitretek Systems, Washington, D.C., July 2002.

9

ITS Challenges and Opportunities

The deployment of ITS has tremendous potential for improving the safety and efficiency of our transportation system. However, for the true benefits of ITS to materialize, a number of challenges will have to be overcome first. These challenges include technical issues, as well as institutional and legal issues. This chapter will discuss some of the most significant challenges, as well as some future opportunities for ITS.

9.1 Mainstreaming ITS

While ITS solutions to transportation problems are now being considered by many transportation agencies all over the country, ITS solutions are still not yet part of the routine metropolitan transportation planning process. Some agencies may view ITS as an element that is too costly to build and may not produce any significant impact in solving any transportation problems. Moreover, several institutional barriers still exist that hinder the full "mainstreaming" of ITS into the regular planning process.

A recent study conduced for the U.S. DOT surveyed 10 metropolitan areas in the United States in an effort to understand the factors that support the consideration of ITS solutions to transportation problems [1]. The study identified the following three main factors that facilitate the routine consideration of ITS:

1. The endorsement of ITS by elected officials and transportation managers, who can then act as ITS champions in the region;

2. Improved communications and coordination across geographic boundaries, since the true benefits of ITS become most obvious when it is deployed with a regional perspective that cuts across geographic boundaries and agencies;
3. The collection and use of ITS-generated data to estimate the benefits and costs of ITS projects, estimate operational costs of ITS systems, and improve the design of future systems.

The study also helped identify a number of strategies that agencies are currently pursuing in order to achieve the aforementioned three conditions. Seven of these strategies are summarized below.

1. *Create an ITS committee that involves regional stakeholders.* Experience shows that such committees help educate members about the benefits of ITS. They also help improve communications across agencies and geographic boundaries.
2. *Educate elected officials and transportation executives.* Educating elected officials and transportation managers about the benefits and potential of ITS technologies is key to mainstreaming ITS.
3. *Include ITS in MPO planning documents.* MPOs are the bodies responsible for transportation planning in metropolitan areas. The inclusion of an ITS solution in MPO documents will help demonstrate to transportation professionals that ITS solutions are being considered seriously as potential solutions to transportation problems. This inclusion could be at the conceptual level (as a part of the regional transportation plan) or at the project level (as a part of the TIP).
4. *Develop a program of regional ITS projects.* As previously mentioned, to be effective, ITS should be deployed with a regional perspective. A regional ITS program can help improve communication and coordination of ITS plans and projects across a defined region. This is especially true if the regional program has the endorsement of key transportation executives or has dedicated funds available to it.
5. *Educate the general public on specific ITS projects.* The success of a number of ITS projects, such as traveler information systems, is heavily dependent on public support and acceptance. Given this, educating the public could be quite helpful in mainstreaming ITS.
6. *Use the National ITS Architecture to develop a regional architecture.* As discussed in Chapter 6, a regional ITS architecture provides a framework for regional ITS planning that encourages coordination among stakeholders and optimizes opportunities for integration. A regional

ITS architecture could serve as an important tool for achieving the conditions needed for mainstreaming ITS.

7. *Determine data collection needs for planning purposes.* Data generated from ITS operations can be useful for a wide range of applications that go beyond their use in the real-time content of ITS. Given this, a wide range of planning functions can be supported with ITS data. This includes the use of data for the development and calibration of travel demand and traffic simulation models, congestion monitoring, transit route and schedule planning, intermodal facilities planning, and air quality modeling. Many agencies around the country have actually started archiving ITS data for use in planning applications. Such efforts can help achieve the third factor previously mentioned.

It should be noted, however, that although the mainstreaming of ITS has yet to occur, there are many positive examples of achievements with ITS deployments. Examples of ITS applications can be found all around the country (see Chapter 4), with several evaluation studies reporting positive and, at many times, significant benefits due to the deployment of ITS (see Chapter 8). Moreover, the deployment of ITS is not just limited to the United States. Today, we see aggressive ITS deployments in many countries around the globe. For example, in-vehicle navigation is a popular option in vehicles sold in Japan. Many Japanese companies share the market with different models of in-vehicle navigation systems. In Japan, millions of cars with in-vehicle navigation systems are already in use. Canada now has many state-of-the-art traffic control centers. Europe has been deploying and operating ITS since the early 1990s.

9.2 System Upgrade

As previously discussed in Chapter 6, the ITS planning process typically starts with the establishment of a vision statement for what the ITS system for the region would look like in the future. In concert with the transportation planning process, a region should perform periodic evaluations to measure their progress in realizing its ITS vision [2]. An ITS migration path should be defined to guide a region toward the vision.

As the world moves through the peak of the information superhighway at the beginning of this century, newer technologies will enable a region to obtain better functionality from the deployed ITS systems. This will bring the need for system upgrades and replacement. An ideal migration path should be sensitive to changes in technologies that will occur over time, so that planned ITS deployments will take into account system upgrade and integration requirements. The migration path should also define the evaluation requirements at specific milestones, so that, if necessary, the vision can be modified.

As new functionalities are sought and new technologies are available, transportation agencies have to embark on the task of system replacement in the future. Thus, system replacement should be a part of any planning for system operations. Conformance to national ITS standards during deployment and the use of open architecture systems to the extent possible are two essential elements that would facilitate system upgrades.

9.3 System Integration

System integration is the task of connecting deployed systems so that optimal functionality is attained. The challenge lies in integrating different systems that start operations at different times using different technologies. The use of proprietary standards and protocols often poses serious problems to system deployment and integration. One solution to this problem would be to require open architecture equipment and the use of standard industry protocols in all projects [2]. This will help reduce the complexity of system integration at the regional, state, and national level. However, realistically, different protocols are being standarized to meet different system requirements. Above all else, if a particular system interface is open (i.e., documentation exists that completely describes the interface), even nonstandardized interfaces can be accessed.

9.4 Training Needs

While technologies will provide greater opportunities to obtain increased functionalities from ITS, it will also require hiring highly technical staff to manage and operate these advanced systems. One of the challenges many pubic agencies may face is that while ITS deployments are calling for hiring professionals with special skills, they are being forced to reduce their workforce due to budgetary constraints. Under this situation, agencies may have to train existing staff to procure, operate, and manage ITS. Another strategy may be to share staffing with participating agencies in any ITS project to minimize the need for additional staffing requirements.

An inflexible agency culture may also cause problems [3]. Agency personnel may be reluctant to accept the new technologies, as well as the new ways of doing business that ITS requires. This could involve modifying the procurement process to accommodate ITS products, more communication with other agencies, and resource sharing.

9.5 Funding

Roadway construction projects provide visible products and benefits to the public. Consequently, these projects are more likely to receive political and public support. Most of the time, however, the presence and benefits of ITS are subtler. For example, an array of sensor systems, a communication system, and a central control center for an ITS project are usually not apparent to road users, and their benefits may not be readily identifiable. This poses a serious challenge to acquire support for ITS projects.

Communicating the benefits of ITS to the public, policy makers, and transportation agencies is the key to bringing ITS into the funding decision scenario. New strategies for making ITS more visible to road users should be developed. One strategy could include using static or dynamic signs to inform the public that ITS is in operation on a particular highway segment. The recent success of Amber Alert messages for informing road users of child kidnappings could increase the desire to have dynamic message signs widely deployed.

9.6 Privacy

The question of whether the privacy issue would constrain ITS deployment has been debated in the transportation industry for more than a decade. Some view it as being a major obstacle in the widespread deployment of ITS, because the public will not accept a technology that may be used as a tool to spy on their life and invade their privacy.

Many perceive the video cameras on highways to monitor the traffic condition as a risk to their privacy. Many residents in several cities object to the deployment of red light running cameras at intersections, as it might compromise their privacy. Similarly, recent technology that allows for potential cellular phone tracking by an information service provider to identify vehicle locations, speeds, and travel time could be perceived as another tool to monitor the public's movements. Selling ITS to the public will be quite challenging if it is still perceived as a serious threat to privacy.

Dealing with the privacy issue is an important matter, and solving it should be seen as a challenge. The operational concept and subsequent system design should address this issue with utmost importance. System planners and designers should minimize and, if possible, eliminate any function that may compromise users' privacy. For example, the video camera in the roadway should be used only to capture the presence of a vehicle without any zooming capability that can identify the vehicle's occupants.

To date, ITS professionals have addressed most of the privacy concerns surrounding ITS. A good example of strategies for addressing privacy concerns is the procedure followed by the University of Washington (UW) in developing the traveler information backbone for the Seattle area [3]. To protect the privacy of individuals, UW staff have developed procedures to remove personal information from the data. Each data source has a firewall to strip out any private data before they go onto the backbone. The stripped data always resides at the source agency. For example, the computer residing at the transit agency would extract bus driver identification before vehicle identification data are passed to the communications backbone.

Another example is found in Phoenix, where transportation officials have tried to counteract concerns about camera use. First, Arizona DOT officials made a linguistic change, replacing the somewhat intrusive-sounding phrase "video surveillance" with the phrase "video monitoring." In addition, the agency personnel agreed that the technology should not play a role in law enforcement, but rather be primarily used for traffic management purposes. Tapes from the camera feeds are even discarded to avoid problems with the tapes being subpoenaed and used in lawsuits [3].

9.7 Rulemaking and Compliance

The TEA-21 included a requirement that U.S. agencies should be consistent with the National ITS Architecture. The objective of this requirement is to ensure that ITS is deployed across the United States using a common framework, so that any future upgrades or additions to ITS can be integrated without any complexity. As discussed in Chapter 5, the National ITS Architecture contains tremendous amounts of information, which could present a challenge to many agencies' personnel that need to go though this information to make sure they are consistent with the architecture. The development of software, such as Turbo Architecture, is aimed at facilitating the application of the National ITS Architecture and will likely help overcome this problem.

9.8 Resource Sharing

Resource-sharing initiatives between public and private agencies is gaining popularity with public agencies interested in aggressive ITS deployment. Under these initiatives, public agencies may, for example, provide right-of-way to a private agency to install fiber-optic cable for landline communication systems or communication towers for wireless communications and, in return, the public agency receives the right to use the same communications channels for free. Resource-sharing initiatives are an excellent way for public agencies to extract funds or free up communications media or devices to support ITS deployments.

Many state agencies in the United States are pursuing various resource-sharing initiatives to support ITS deployments.

9.9 ITS and National Security

Since the inception of the ITS program, several ITS applications have been designed to improve the safety and security of the transportation system. The tragic events of September 11, 2001, however, have raised the nation's consciousness about the need for better infrastructure protection and more effective detection and responses to natural and man-made disasters. In 2002, the Intelligent Transportation Systems Society of America (ITS America) published a document titled "Homeland Security and ITS" that describes the role that ITS can play in enhancing the surface transportation aspects of homeland security [4]. This document suggests the addition of a security-related goal to the national ITS program plan. That goal is to provide a transportation system that does the following:

- Protects well against attacks;
- Responds rapidly and effectively to natural and human-caused threats and disasters;
- Supports appropriate transportation and emergency management agencies;
- Moves people effectively, even during a crisis;
- Can be restored quickly to its full capacity.

The ITS America document discusses five broad areas where ITS can contribute to enhancing homeland security:

1. *Preparedness.* This area focuses on understanding where the existing vulnerabilities in the transportation system are and identifying existing technologies that can help reduce such vulnerabilities. It also focuses on developing tools and technologies to facilitate communications and coordination among the several agencies that would be involved in a time of crisis.
2. *Prevention.* Prevention refers to counteracting threats before they take place. The goal is to develop technologies that can detect and head off attacks along rails and roads. Examples of such technologies would include the following:

- Technologies against the misuse of commercial vehicles that could prevent the deviation of the vehicle from its preplanned route;
- Surveillance technologies to protect the road and rail infrastructure from tampering;
- Pattern recognition technologies.

3. *Protection.* Protection refers to preventing attacks and minimizing their consequences. To achieve this, this area focuses on developing tools and technology for on-site detection and response to threats to facilities and infrastructure systems, activating alternate routes during a crisis, and increasing the ability of agencies to undertake protective activities.

4. *Response.* The goal here is to support responding agencies and increase their effectiveness. To do this, this area strives to develop technologies for the following:
   - Maintain communications among responding agencies;
   - Disseminate real-time information about the status of the transportation system;
   - Track the location of vehicles carrying hazardous materials that are close to a crisis scene;
   - Reroute traffic when parts of the transportation system are impaired.

5. *Recovery.* This involves the ability to restore the system to its original capacity after a disaster. Efforts in this area will focus on creating a flexible, reconfigurable transportation system that can meet the needs of emergency situations. They will also involve using technology to allow for effectively executing plans for alternate routes and modes during a crisis and for making the best use of available capacity.

It is anticipated that the next major revision of the National ITS Architecture will contain substantial security additions. Hopefully, these potential security areas will spur the ITS community to aid homeland security and secure ITS.

9.10 Conclusions

In this chapter, we briefly discussed some of ITS challenges as well as opportunities. Many efforts, resources, and research have been expended to date to bring the benefits of ITS to solve our transportation problems for today and, more importantly, for the future. This is likely to continue into the future, as we try to learn from these early deployments in order to improve the effectiveness of ITS in improving the efficiency and safety of the transportation system.

Review Questions

1. What are the different approaches to mainstream ITS into general transportation decision making and activities?
2. What are the challenges related to funding for ITS projects? How can these challenges be met?
3. Will rulemaking by U.S. DOT for conformance with the National ITS Architecture help mainstream ITS into the ove. ll transportation framework? Discuss your opinion.
4. How will ITS programs improve the safety and security of the transportation system? Discuss other ways (that are not discussed in the text) that ITS could be applied to improve transportation security.

References

[1] Deysher, E., D. W. Jackson, and A. J. DeBlasio, "Incorporating ITS Solutions into the Metropolitan Transportation Planning Process," Prepared for the U.S. Department of Transportation, available from the ITS Electronic Document Library (EDL) at http://www.its.dot.gov/welcome.htm, EDL Document Number: 13177, 2000.

[2] Virgina Department of Transportation, *Smart Travel Business Plan, 1997–2006.*

[3] U.S. Department of Transportation, Federal Highway Administration, "What Have We Learned About Intelligent Transportation Systems," available from the ITS Electronic Document Library (EDL) at http://www.its.dot.gov/welcome.htm, 2000.

[4] Intelligent Transportation Society of America, "ITS and Homeland Security. Using Intelligent Transportation Systems to Improve and Support Homeland Security," supplement to the "National ITS Program Plan: A Ten-Year Vision," Washington, D.C., 2002.

About the Authors

Mashrur A. Chowdhury is an assistant professor in the Department of Civil and Environmental Engineering and Engineering Mechanics at the University of Dayton in Dayton, Ohio. He received his Ph.D. in civil engineering from the University of Virginia. Dr. Chowdhury teaches graduate and undergraduate classes in intelligent transportation systems (ITS), transportation systems management, and transportation engineering. He has been involved as a principal investigator in research related to ITS, transportation investments, and operations funded by the U.S. Department of Transportation, the Ohio Department of Transportation, and the Ohio Board of Regents. Dr. Chowdhury has published his research results in such journals as *American Society of Civil Engineers Journal on Infrastructure Systems, Reliability Engineering and System Safety,* and *The Institute of Transportation Engineers (ITE) Journal.*

Dr. Chowdhury is a registered professional engineer in Washington, D.C. Previously, he worked as senior engineer with BMI, Viggen Corporations, and Iteris, where he worked on numerous ITS-related projects. He is a member of the Transportation Research Board's committee on artificial intelligence and the American Society of Civil Engineers' (ASCE) committee on computing in transportation. Dr. Chowdhury is a past chair of the ASCE committee on computing in transportation.

Adel Sadek is an assistant professor of civil engineering at the University of Vermont, where he teaches and conducts research in the areas of ITS, computational intelligence applications in transportation, and transportation systems modeling and analysis. Dr. Sadek received his Ph.D. from the University of Virginia. His dissertation, titled "Case-Based Reasoning for Real-Time Traffic Flow Management," won the 1998 Milton Pikarsky Award for the best dissertation in the field of transportation science and technology awarded by the Council of University Transportation Centers.

Dr. Sadek's research has been funded by several agencies, including the National Science Foundation, the New England University Transportation Center, the New England Transportation Consortium, and the Vermont Agency of Transportation. He has published his research results in such journals as *Transportation Research, The Journal of the Transportation Research Board, Computer-Aided Civil and Infrastructure Engineering Journal,* and *The Journal of Transportation Engineering.* Dr. Sadek is a member of the Transportation Research Board's committee on artificial intelligence and a member of the ASCE's committee on computing in transportation.

Index

Actuated signals, 21–24
    control operational concept, 22
    defined, 18
    maximum green, 21
    minimum green, 21
    parameters, 21
    passage time interval, 21
    See also Traffic signals
Adaptive traffic control systems, 75–80
    algorithms, 76–78
    computer traffic control systems and, 75–76
    defined, 75
    real-world, 78–80
    RHODES, 78, 79
    SCATS, 77–78, 80
    SCOOT, 77, 80
    See also ITS applications
Advanced public transportation systems, 80–86
    AVL systems, 80–81
    electronic fare payment systems, 86
    transit information systems, 83–86
    transit operations software, 82–83
    See also ITS applications
Advanced Rural Transportation System (ARTS), 55
Advanced vehicle control and safety systems bundle, 48–52
Archived Data User Service (ADUS), 52

Automated highway system (AHS), 52
Automated roadside safety inspection, 45
Automatic incident detection (AID) systems, 68
Automatic vehicle identification (AVI), 63–64
Automatic vehicle location (AVL), 63–64, 80–81
    defined, 80
    real-world implementations, 81
    See also Advanced public transportation systems

Benefit-cost analysis, 154

Car-following models, 14
Cellular phones
    hotlines, 73
    in incident detection, 65
Center-to-center (C2C) applications, 141–44
    data archival, 142
    incident management, 142
    rail coordination, 142
    traffic management, 143
    transit management, 143–44
    traveler information, 144
    See also ITS standards
Center-to-roadside interface class, 138–41
    data collection/monitoring, 138–40
    dynamic message signing, 140
    environmental monitoring, 140

Center-to-roadside interface class (continued)
  ramp metering, 140
  traffic signal, 140–41
  vehicle sensors, 141
  video surveillance, 141
  See also ITS standards
Center-to-vehicle/traveler applications, 144–45
  mayday, 144
  transit vehicle communications, 144–45
  traveler information, 145
  See also ITS standards
Closed-circuit TV (CCTV), 63
Coalition building, 4
Commercial radio, 74
Commercial vehicle administrative processes, 46
Commercial vehicle electronic clearance, 45
Commercial vehicles operations bundle, 45–47
Computer-aided dispatching (CAD), 82
Computer traffic control systems, 75–76
Concept of operations, 95–97
  defined, 95–96
  illustrated, 97
  for regional ITS architecture, 96
Congestion
  nonrecurrent, 56, 67
  recurrent, 56, 67
Congestion Management System (CMS), 120
Contract management, 4
Coordinated systems, 19
Corsim model, 31–32, 166–67
Crossing left turn (CLT) collisions, 50
Cycles, 19

Data
  analysis/design, 3
  archival, 142
  collection, 59–60, 138–40
  monitoring, 138–40
Data flow diagrams (DFDs), 102
Decision support systems (DSSs)
  incident response, 69
  in management/control decisions, 57
Dedicated short-range communications (DSRC), 105
Demand-capacity control, 29

Density
  defined, 8
  flow relationship, 8, 9
  speed relationship, 9, 10
Deployment tracking, 155–60
  defined, 155
  integration, 155–59
  progress, 160
  See also ITS evaluation
Detection methods, 61–65
  acoustic detectors, 62
  CCTV, 63
  environmental sensors, 65
  ILDs, 61
  infrared sensors, 62
  microwave radar detectors, 61–62
  mobile reports, 64–65
  ultrasonic detectors, 62
  vehicle probes, 63–64
  VIP, 63
  See also Traffic surveillance
Deterministic models, 31
Display terminals, 74
Dynamic message signs (DMS), 72–73, 140
  hybrid, 73
  light-emitting, 72
  light-reflecting, 72
Dynamit model, 166

Electronic fare payment systems, 86
Electronic payment bundle, 44–45
Electronic payment services (EPS), 44–45
  components, 44–45
  defined, 44
  integrated, 44–45
Electronic toll collection (ETC), 44
Emergency management bundle, 47–48
Emergency notification and personal security, 47
Emergency vehicle management, 48
Emissions testing and mitigation, 41
En route driver information, 38–39
En route transit information, 43
Entrance ramp closure, 25
Entrance ramp metering, 25
Environmental monitoring, 140
Environmental sensors, 65
Equipment packages, 107
Evaluation. *See* ITS evaluation

Exit ramp closure, 25
Experience-based testing, 150

Financing, 4
Flow
    defined, 7
    density relationship, 8, 9
    speed relationship, 9, 10
Freeway and incident management systems
        (FIMS), 56–74
    benefits, 74
    defined, 56
    functions, 57–59
    incident management, 57–58, 67–71
    information dissemination, 58, 71–74
    lane use control, 58
    objectives, 56–57
    preferential HOV treatment, 59
    ramp control, 57, 66–67
    traffic surveillance and incident detection,
        57, 59–66
    See also ITS applications
Freeway Management Handbook, 56
Freeway service patrol, 65
Freight mobility, 47
Fresim, 31
Funding, 175

Geographic information systems (GIS), 70
Global positioning systems (GPS), 64
Greenberg model, 13
Greenshields model, 11–13
    defined, 11
    equation, 11
    See also Traffic flow models

Hazardous materials incident response, 47
Headway, 8
High-occupancy vehicles (HOVs), 40, 41
    facility management and control, 41
    preferential treatment to, 59
Highway advisory radio (HAR), 73
Highway capacity manual (HCM), 21
Highway capacity software (HCS), 21
Highway-rail intersection (HRI), 41–42
    application area, 146–47
    equipment health, 42
    warning/control devices, 41

Impact assessment, 160–62
    effectiveness measures, 160, 161

physical contexts, 160
    See also ITS evaluation
Incident detection, 57
Incident management, 40
    application area, 142
    clearance stage, 70–71
    defined, 57–58, 67
    detection and verification stage, 68
    goal, 67, 68
    recovery stage, 71
    response stage, 68–70
    stages, 67
Incident response, 68–70
    alternative route planning, 70
    defined, 68–69
    DSSs, 69
    emergency vehicle access, 69–70
    freeway service patrols, 70
    goal, 69
    manuals, 69
    tow truck contracts, 69
    traffic management/control, 70
    See also Incident management
Inductive loop detectors (ILDs), 61
Information dissemination, 71–74
    defined, 58, 71
    effectiveness, 71–72
    goal, 71
    in-vehicle devices, 73–74
    off-roadway devices, 74
    on-roadway devices, 72
Information management bundle, 52
Infrared sensors, 62
Integration model, 32, 166
Intelligent Transportation System
        Deployment Analysis System
        (IDAS), 132
Intelligent Transportation Systems. *See* ITS
Interchangeability, 136
In-terminal/wayside transit information
        systems, 84–85
Interoperability
    importance, 94
    ITS standards support, 135–36
    national, 94
Intersections
    collision avoidance, 50
    coordinated system, 19
    isolated, 19

Interval, 19
In-vehicle information devices, 73–74
    commercial radio, 74
    display terminals, 73
    HAR, 73
    hotlines, 73
    See also Information dissemination
In-vehicle transit information systems, 85–86
ITS
    components, benefits of, 162
    components and elements, 163
    defined, 1
    interoperability, 94
    mainstreaming, 171–73
    migration path, 173–74
    national security and, 177–79
    potential, 1–2
    strategies, 172–73
    today and tomorrow, 2
    training/education needs, 2–4
    transportation planning and, 113–17
    transportation planning integration, 125–32
ITS applications, 2, 55–89
    adaptive traffic control systems, 75–80
    advanced public transportation systems, 80–86
    FIMS, 56–74
    multimodal traveler information systems, 86–89
ITS architecture, 93–111
    concept of operations, 95–97
    design vs., 93
    national, 98–108
    need for, 94–95
    project, 93–94
    regional, 93, 95–96, 108
ITS challenges, 171–79
    funding, 175
    mainstreaming, 171–73
    national security, 177–78
    privacy, 175–76
    resource sharing, 176–77
    rulemaking/compliance, 176
    system integration, 174
    system upgrade, 173–74
    training needs, 174
ITS Deployment Analysis System, 165–66
    availability, 165–66
    components, 166
    defined, 165
    interfaced, 166
ITS evaluation, 153–68
    benefit estimation categories, 162–63
    categories and tools, 165
    challenges, 167–68
    deployment tracking, 155–60
    guidelines, 163–65
    IDAS, 165–66
    impact assessment, 160–62
    role of, 154
    support tools, 165–67
    TEA-21 guidelines, 165
    traffic simulation models and, 166–67
    types, 153
ITS planning, 119–23
    approaches, 120
    case study, 124–25
    market package-based, 121
    National ITS Architecture and, 117–19
    project selection, 154–55
    traceability-based, 121–23
    training/education and, 3
    transportation planning vs., 115
ITS standards, 135–51
    application areas, 138–47
    center-to-center (C2C) applications, 141–44
    center-to-roadside interface class, 138–41
    center-to-vehicle/traveler applications, 144–45
    development process, 136–37
    introduction, 135–36
    National ITS Architecture and, 137–38
    NTCIP, 147–50
    roadside-to-roadside applications, 146–47
    roadside-to-vehicle applications, 145–46
    support, 135–36
    testing, 150
ITS user services, 35–53
    advanced vehicle control and safety systems bundle, 48–52
    automated highway system, 52
    automated roadside safety inspection, 45
    bundles, 36
    characteristics, 35–36
    commercial vehicle administrative processes, 46

commercial vehicle electronic clearance, 45
commercial vehicles operations bundle, 45–47
electronic payment bundle, 44–45
electronic payment services (EPS), 44–45
emergency management bundle, 47–48
emergency notification and personal security, 47
emergency vehicle management, 48
emissions testing and mitigation, 41
en route driver information, 38–39
en route transit information, 43
freight mobility, 47
hazardous materials incident response, 47
highway-rail intersection, 41–42
incident management, 40
information management bundle, 52
intersection collision avoidance, 50
lateral collision avoidance, 49–50
list of, 37
longitudinal collision avoidance, 48–49
maintenance and construction operations (MCO), 52–53
onboard safety monitoring, 45–46
personalized public transit (PPT), 43
precrash restraint deployment, 51
pretrip travel information, 38
public transportation management, 42–43
public transportation operations bundle, 42–44
public travel safety, 44
ride matching and reservation, 39
route guidance, 39
safety readiness, 51
traffic control, 40
travel and traffic management bundle, 36–42
travel demand management, 40–41
traveler services information, 39
vision enhancement for collision avoidance, 51

Lane use control, 58
Lateral collision avoidance, 49–50
 defined, 49
 subservices, 49–50
 See also ITS user services
Local-area networks (LANs), 66
Logical architecture, 101–4

 components, 102
 defined, 101
 highest level of, 103
 implementation details, 102
 moving to physical architecture, 104
 See also national ITS architecture
Longitudinal collision avoidance, 48–49
 defined, 48
 systems, 48, 49
 See also ITS user services

Macroscopic models, 31
Mainstreaming ITS, 171–73
Maintenance and construction operations (MCO), 52–53
Market package-based ITS planning, 121
 defined, 121
 illustrated, 122
 vision/goals, 121
 See also ITS planning
Market packages, 107–8
Mayday, 144
Mesoscopic models, 31
Microscopic models, 14, 31
Microwave radar detectors, 61–62
Mobile reports, 64–65
MPOs, 172
Multimodal traveler information systems, 86–89
 benefits, 89
 defined, 86–87
 potential contents, 87
 Smart Trek (Seattle), 87–89
 See also ITS applications

National ITS Architecture, 98–108
 data set definition, 117–18
 equipment packages, 107
 facilitating application of, 109
 logical architecture, 101–4
 market packages, 107–8
 NTCIP in, 148
 objectives, 147
 physical architecture, 104–6
 planning and, 117–19
 standards and, 137–38
 subsystems and communications, 106
 user services/user service requirements, 98–101
 See also ITS architecture

National ITS Program Plan, 99
    defined, 99
    editing, 101
National security (ITS), 177–79
    enhancement, 177–78
    goals, 177
    preparedness, 177
    prevention, 177–78
    protection, 178
    recovery, 178
    response, 178
National Transportation Communications for ITS Protocol. *See* NTCIP
Netsim, 31
Nonrecurrent congestion, 56, 67
Northwestern model, 13
NTCIP, 105, 147–50
    defined, 147
    in diverse ITS areas, 149
    ITS application areas, 147–50
    in National ITS Architecture, 148

Occupancy control, 29
Offset, 19
Onboard safety monitoring, 45–46
Organization, this book, 5–6

Personal digital assistants (PDAs), 74
Personalized public transit (PPT), 43
Phase, 19
Physical architecture, 104–6
    defined, 104
    illustrated, 106
    interfaces, 105
    moving from logical to, 104
    See also National ITS Architecture
Planning. *See* ITS planning; Transportation planning
Platoon metering, 27–28
Precrash restraint deployment, 51
Pretimed metering, 25–28
    defined, 25–26
    illustrated, 26
    system layout, 26–27, 29
    system operation, 27–28
    See also Ramp metering
Pretrip transit information systems, 83–84
Pretrip travel information, 38
Privacy, 175–76
Project selection, 154–55

benefit-cost analysis, 154
relative rating/ranking, 155
Public transportation management, 42–43
Public transportation operations bundle, 42–44
Public travel safety, 44

Queue formation
    illustrated, 17
    length determination, 17–18

Rail coordination, 142
Ramp metering, 24–30, 66–67
    application area, 140
    defined, 24, 66–67
    entrance, 25
    objectives, 24
    operational concept, 25–30
    platoon, 27–28
    pretimed, 25–28
    single-entry, 27
    system, 29–30
    system use, 24, 66–67
    traffic responsive, 28–29
    two-abreast, 28
    types of, 25
Recovery shock waves, 17
Recurrent congestion, 56, 67
Regional ITS architecture, 93
    concept of operations, 96
    developing, 95
    development procedure, 108
    stakeholders, 96
    See also ITS architecture
Relative rating/ranking, 155
Resource sharing, 176–77
RHODES, 78, 79
Ride matching and reservation, 39
Roadside-to-roadside applications, 146–47
Roadside-to-vehicle applications, 145–46
    signal priority, 146
    toll/fee collection, 145–46
    See also ITS standards
Route guidance, 39
Rulemaking/compliance, 176

Safety readiness, 51
SCATS, 77–78, 80
SCOOT, 77, 80
Scope, this book, 4–5

Shock waves
   analysis, 16
   analysis example, 16–18
   defined, 14
   recovery, 17
   at signalized intersections, 14
   in traffic streams, 14–18
   velocity, 15–16
Signal coordination, 22–24
   common cycle requirement, 23
   defined, 22
   SYNCHRO, 24
   TRANSYT-7F, 23–24
   See also Traffic signals
Signal priority, 146
Single-entry metering, 27
Single-occupancy vehicles (SOVs), 40, 41
Speed
   defined, 8
   density relationship, 9, 10
   flow relationship, 9, 10
Stakeholders, regional ITS architecture, 96
Standard Development Organizations (SDOs), 136–37
Statewide transportation plan, 115, 118
Stochastic models, 30–31
Straight crossing path (SCP) collisions, 50
Survey questionnaire, 156–59
SYNCHRO, 24
System analysis/design, 3
System integration, 174
System metering, 29–30
   defined, 29
   rates computation, 30
   system layout, 30
   See also Ramp metering
Systems integration, 3, 95
System upgrade, 173–74

Technology evaluation, 3
Testing, standards, 150
Toll/fee collection, 145–46
Traceability-based ITS planning, 121–23
   defined, 121–23
   deployment, 123
   framework illustration, 123
   functions, 123
   See also ITS planning
Traceability goals
   "enhance agency operations," 128
   "enhance mobility," 131
   "enhance public safety," 126
   "operate effectively and efficiently," 127
   "support traveler information services," 129
   "user-friendly transportation system," 130
Traffic control, 40
Traffic flow
   elements, 7–11
   fundamental diagram, 9–11
Traffic flow models, 11–14
   alternative, 13–14
   car-following, 14
   Greenberg, 13
   Greenshields, 11–13
   microscopic flow, 14
   Northwestern, 13
   single-regime, 13
   three-regime linear, 13
   two-regime linear, 13
   Underwood, 13
Traffic management application area, 143
Traffic management center (TMC), 41
Traffic responsive metering, 28–29
   defined, 28
   strategy, 28–29
   system layout, 29
   See also Ramp metering
Traffic signalization principles, 18–21
Traffic signals
   actuated, 18, 21–22
   application area, 140–41
   computer-based control, 20
   coordination, 22–24
   cycles, 19
   full-actuated, 20
   modes of operation, 19–20
   pretimed, 18, 19
   priority, 146
   role, 18
   semiactuated, 19–20
   timing principles, 20–21
Traffic simulation models, 30–32
   application guidelines, 32
   classification schemes, 30–31
   Corsim, 31–32, 166–67
   deterministic, 31
   Dynamit, 166
   examples, 31–32

Traffic simulation models (continued)
  Integration, 32, 166
  macroscopic, 31
  mesoscopic, 31
  microscopic, 31
  stochastic, 30–31
Traffic streams, shock waves in, 14–18
Traffic surveillance
  communications system, 66
  data collection, 59–60
  defined, 57
  detection methods, 61–65
  hardware, 65–66
  purposes, 59
  software, 66
  system components, 60
Training needs, 174
Transit information systems, 83–86
  in-terminal/wayside, 84–85
  in-vehicle, 85–86
  pretrip, 83–84
  See also Advanced public transportation systems
Transit management application area, 143–44
Transit operations software, 82–83
  defined, 82
  demand responsive transit operations, 82–83
  fixed-route bus operations, 82
  See also Advanced public transportation systems
Transit vehicle communications, 144–45
Transportation Analysis and Simulation System (TRANSIMS), 132
Transportation Efficiency Act for the Twenty-first Century (TEA-21), 136–37, 165

Transportation Improvement Program (TIP), 113
Transportation planning
  ITS and, 113–17
  ITS integration, 125–32
  ITS planning process vs., 115
  in project life cycle, 120
  statewide, 115, 118
  urban, 114, 119
TRANSYT, 23–24
  defined, 23
  objective functions, 24
  signal plan optimization function, 23–24
  traffic flow simulation function, 23
  See also Signal coordination
Travel and traffic management bundle, 36–42
Travel demand management, 40–41
Traveler information, 144, 145
Traveler services information, 39
Turbo Architecture, 109, 110
Two-abreast metering, 28

Ultrasonic detectors, 62
Underwood model, 13
Urban transportation plan, 119

Validation testing, 150
Variable message signs (VMSs), 38
VDOT NOVA case study, 124–25
Vehicle sensors, 141
Verification testing, 150
Video image processing (VIP), 63
Video surveillance, 141
Vision enhancement for collision avoidance, 51

Writing/communications, 4

Recent Titles in the Artech House ITS Library

John Walker, Series Editor

Advanced Traveler Information Systems, Bob McQueen, Rick Schuman, and Kan Chen

Advances in Mobile Information Systems, John Walker, editor

Fundamentals of Intelligent Transport Systems Planning Mashrur A. Chowdhury and Adel Sadek

Incident Management in Intelligent Transportation Systems, Kaan Ozbay and Pushkin Kachroo

Intelligent Transportation Systems Architectures, Bob McQueen and Judy McQueen

Introduction to Transportation Systems, Joseph Sussman

ITS Handbook 2000: Recommendations from the World Road Association (PIARC), PIARC Committee on Intelligent Transport (Edited by Kan Chen and John C. Miles)

Positioning Systems in Intelligent Transportation Systems, Chris Drane and Chris Rizos

Sensor Technologies and Data Requirements for ITS, Lawrence A. Klein

Smart Highways, Smart Cars, Richard Whelan

Tomorrow's Transportation: Changing Cities, Economies, and Lives, William L. Garrison and Jerry D. Ward

Vehicle Location and Navigation Systems, Yilin Zhao

Wireless Communications for Intelligent Transportation Systems, Scott D. Elliott and Daniel J. Dailey

For further information on these and other Artech House titles, including previously considered out-of-print books now available through our In-Print-Forever® (IPF®) program, contact:

Artech House
685 Canton Street
Norwood, MA 02062
Phone: 781-769-9750
Fax: 781-769-6334
e-mail: artech@artechhouse.com

Artech House
46 Gillingham Street
London SW1V 1AH UK
Phone: +44 (0)20 7596-8750
Fax: +44 (0)20 7630 0166
e-mail: artech-uk@artechhouse.com

Find us on the World Wide Web at:
www.artechhouse.com